T0353364

Mathematical Olympiad in China (2015–2016)

Problems and Solutions

Mathematical Olympiad Series

ISSN: 1793-8570

Series Editors: Lee Peng Yee *(Nanyang Technological University, Singapore)*
Xiong Bin *(East China Normal University, China)*

Published

Vol. 19 *Mathematical Olympiad in China (2019–2020):*
Problems and Solutions
edited by Bin Xiong (East China Normal University, China)

Vol. 18 *Mathematical Olympiad in China (2017–2018):*
Problems and Solutions
edited by Bin Xiong (East China Normal University, China)

Vol. 17 *Mathematical Olympiad in China (2015–2016):*
Problems and Solutions
edited by Bin Xiong (East China Normal University, China)

Vol. 16 *Sequences and Mathematical Induction:*
In Mathematical Olympiad and Competitions
Second Edition
by Zhigang Feng (Shanghai Senior High School, China)
translated by: Feng Ma, Youren Wang

Vol. 15 *Mathematical Olympiad in China (2011–2014):*
Problems and Solutions
edited by Bin Xiong (East China Normal University, China) &
Peng Yee Lee (Nanyang Technological University, Singapore)

Vol. 14 *Probability and Expectation*
by Zun Shan (Nanjing Normal University, China)
translated by: Shanping Wang (East China Normal University, China)

Vol. 13 *Combinatorial Extremization*
by Yuefeng Feng (Shenzhen Senior High School, China)

Vol. 12 *Geometric Inequalities*
by Gangsong Leng (Shanghai University, China)
translated by: Yongming Liu (East China Normal University, China)

The complete list of the published volumes in the series can be found at
http://www.worldscientific.com/series/mos

Vol. 17 | Mathematical Olympiad Series

Mathematical Olympiad

in China (2015–2016)

Problems and Solutions

Editor-in-Chief
Xiong Bin
East China Normal University, China

English Translator
Chen Xiao-Min
Shanghai Hypers Inc. China

Copy Editors
Ni Ming
Kong Ling-Zhi
East China Normal University Press, China

East China Normal
University Press

World Scientific

Published by

East China Normal University Press
3663 North Zhongshan Road
Shanghai 200062
China

and

World Scientific Publishing Co. Pte. Ltd.
5 Toh Tuck Link, Singapore 596224
USA office: 27 Warren Street, Suite 401-402, Hackensack, NJ 07601
UK office: 57 Shelton Street, Covent Garden, London WC2H 9HE

Library of Congress Cataloging-in-Publication Data
Names: Xiong, Bin, editor.
Title: Mathematical Olympiad in China (2015–2016) : problems and solutions /
    editor Xiong Bin, East China Normal University, China.
Description: Shanghai, China : East China Normal University Press ; Hackensack, NJ :
    World Scientific, [2022] | Series: Mathematical Olympiad series, 1793-8570 ; vol. 17
Identifiers: LCCN 2022018127 | ISBN 9789811250712 (hardcover) |
    ISBN 9789811251917 (paperback) | ISBN 9789811250729 (ebook) |
    ISBN 9789811250736 (ebook other)
Subjects: LCSH: International Mathematical Olympiad. | Mathematics--Problems, exercises, etc. |
    Mathematics--Competitions--China.
Classification: LCC QA43 .M314537 2022 | DDC 510.76--dc23/eng/20220330
LC record available at https://lccn.loc.gov/2022018127

British Library Cataloguing-in-Publication Data
A catalogue record for this book is available from the British Library.

For any available supplementary material, please visit
https://www.worldscientific.com/worldscibooks/10.1142/12681#t=suppl

Typeset by Stallion Press
Email: enquiries@stallionpress.com

Printed in Singapore

Preface

The first time China participate in IMO was in 1985 where two students were sent to the 26th IMO. Since 1986, China has a team of 6 students at every IMO except in 1998 when it was held in Taiwan. So far, up to 2016, China has achieved the number one ranking in team effort 22 times. A great majority of students received gold medals. The fact that China obtained such encouraging result is due to, on one hand, Chinese students' hard work and perseverance, and on the other hand, the effort of the teachers in schools and the training offered by national coaches. We believe this is also a result of the education system in China, in particular, the emphasis on training of the basic skills in science education.

The materials of this book come from a series of two books (in Chinese) on *Forward to IMO: A collection of Mathematical Olympiad Problems* (2015–2016). It is a collection of problems and solutions of the major mathematical competitions in China. It provides a glimpse of how the China national team is selected and formed. First, there is the China Mathematical Competition, a national event. It is held on the second Sunday of September every year. Through the competition, about 350 students are selected to join the China Mathematical Olympiad (commonly known as the winter camp), or in short CMO, in December. CMO lasts for five days. Both the type and the difficulty of the problems match those of IMO. Similarly, students are given three problems to solve in 4.5 hours each day. From CMO, 60 students are selected to form a national training team. The training takes place for two weeks in the month of March. After four to six tests, plus two qualifying examinations, six students are finally selected to form the national team, taking part in IMO in July of that year.

In view of the differences in education, culture and economy of the western part of China and with the coastal part in eastern China, mathematical competitions in West China did not develop as fast as the rest of the country. In order to promote the activity of mathematical competition, and to enhance the level of mathematical competition, starting from 2001, China Mathematical Olympiad Committee organizes the China Western Mathematical Olympiad.

Since 2012, the China Western Mathematical Olympiad has been renamed the China Western Mathematical Invitation. The competition dates have been changed from the first half of October to the middle of August since 2013.

The development of this competition reignited the enthusiasm of Western students for mathematics. Once again, the figure of Western students often appeared in the national team.

Since 1995, there was no female student in the Chinese national team. In order to encourage more female students participating in the mathematical competition, starting from 2002, China Mathematical Olympiad Committee conducted the China Girls' Mathematical Olympiad. Again, the top twelve winners will be admitted directly into the CMO.

The authors of this book are coaches of the China national team. They are Xiong Bin, Wu Jianping, Leng Gangsong, Yu Hongbing, Li Weigu, Zhu Huawei, Li Shenghong, Yao Yijun, Feng Zhigang, Qu Zhenhua, Li Qiusheng, He Yijie, Zhang Sihui and Lin Tianqi. The translator of this book is Chen Xiaomin. We are grateful to Qiu Zonghu, Wang Jie, Zhou Qing, Wu Jianping, and Pan Chengbiao for their guidance and assistance to the authors. We are grateful to Ni Ming, Kong Lingzhi of East China Normal University Press. Their effort has helped make our job easier. We are also grateful to Zhang Ji of World Scientific Publishing for her hard work leading to the final publication of the book.

Authors

October 2016

Introduction

Early days

The International Mathematical Olympiad (IMO), founded in 1959, is one of the most competitive and highly intellectual activities in the world for high school students.

Even before IMO, there were already many countries which had mathematics competition. They were mainly the countries in Eastern Europe and in Asia. In addition to the popularization of mathematics and the convergence in educational systems among different countries, the success of mathematical competitions at the national level provided a foundation for the setting-up of IMO. The countries that asserted great influence are Hungary, the former Soviet Union and the United States. Here is a brief history of the IMO and mathematical competition in China.

In 1894, the Department of Education in Hungary passed a motion and decided to conduct a mathematical competition for the secondary schools. The well-known scientist, *J. von Etövös*, was the Minister of Education at that time. His support in the event had made it a success and thus it was well publicized. In addition, the success of his son, *R. von Etövös*, who was also a physicist, in proving the principle of equivalence of the general theory of relativity by *A. Einstein* through experiment, had brought Hungary to the world stage in science. Thereafter, the prize for mathematics competition in Hungary was named "*Etövös* prize". This was the first formally organized mathematical competition in the world. In what follows, Hungary had indeed produced a lot of well-known scientists including *L. Fejér*, *G. Szegö*, *T. Radó*, *A. Haar* and *M. Riesz* (in real analysis), *D. König* (in combinatorics), *T. von Kármán* (in aerodynamics), and *J. C. Harsanyi* (in game theory), who had also won the Nobel Prize for Economics in 1994. They all were the winners of Hungary mathematical competition. The top scientific genius of Hungary, *J. von Neumann*, was one of the leading mathematicians in the 20th century. *Neumann* was overseas while the competition took place. Later he did the competition himself and it took him half an hour to complete. Another mathematician worth mentioning is the highly productive number theorist *P. Erdős*. He was a pupil of *Fejér* and also a winner of the Wolf Prize. *Erdős* was very passionate about mathematical competition and setting competition questions. His contribution to discrete mathematics was unique and greatly significant. The rapid progress and development of discrete mathematics over the subsequent decades had indirectly influenced the types of questions set in IMO. An internationally recognized prize was named after *Erdős* to honour those who had contributed to the education of mathematical competition. Professor *Qiu Zonghu* from China had won the prize in 1993.

In 1934, a famous mathematician *B. Delone* conducted a mathematical competition for high school students in Leningrad (now St. Petersburg). In 1935, Moscow also started organizing such events. Other than being interrupted during the World War II, these events had been carried on until today. As for the Russian Mathematical Competition (later renamed as the Soviet Mathematical Competition), it was not started until 1961. Thus, the former Soviet Union and Russia became the leading powers of Mathematical Olympiad. A lot of grandmasters in mathematics including the great *A. N. Kolmogorov* were all very enthusiastic about the mathematical competition. They would personally involve themselves in setting the questions for the competition. The former Soviet Union even called it the Mathematical Olympiad, believing that mathematics is the "gymnastics of thinking". These points of view gave a great impact on the educational community. The winner of the Fields Medal in 1998, *M. Kontsevich*, was once the first runner-up of the Russian Mathematical Competition. *G. Kasparov*, the international chess grandmaster, was once the second runner-up. *Grigori Perelman*, the winner of the Fields Medal in 2006 (but he declined), who solved the Poincaré's Conjecture, was a gold medalist of IMO in 1982.

In the United States of America, due to the active promotion by the renowned mathematician *G.D. Birkhoff* and his son, together with *G. Pólya*, the Putnam mathematics competition was organized in 1938 for junior undergraduates. Many of the questions were within the scope of high school students. The top five contestants of the Putnam mathematical competition would be entitled to the membership of Putnam. Many of these were eventually outstanding mathematicians. There were the famous *R. Feynman* (winner of the Nobel Prize for Physics, 1965), *K. Wilson* (winner of the Nobel Prize for Physics, 1982), *J. Milnor* (winner of the

Fields Medal, 1962), *D. Mumford* (winner of the Fields Medal, 1974), and *D. Quillen* (winner of the Fields Medal, 1978).

Since 1972, in order to prepare for the IMO, the United States of America Mathematical Olympiad (USAMO) was organized. The standard of questions posed was very high, parallel to that of the Winter Camp in China. Prior to this, the United States had organized American High School Mathematics Examination (AHSME) for the high school students since 1950. This was at the junior level and yet the most popular mathematics competition in America. Originally, it was planned to select about 100 contestants from AHSME to participate in USAMO. However, due to the discrepancy in the level of difficulty between the two competitions and other restrictions, from 1983 onwards, an intermediate level of competition, namely, American Invitational Mathematics Examination (AIME), was introduced. Henceforth both AHSME and AIME became internationally well-known. Since 2000, AHSME was replaced by AMC 12 and AMC 10. Students who perform well on the AMC 12 and AMC 10 are invited to participate in AIME. The combined scores of the AMC 12 and the AIME are used to determine approximately 270 individuals that will be invited back to take the USAMO, while the combined scores of the AMC 10 and the AIME are used to determine approximately 230 individuals that will be invited to take the USAJMO (United States of America Junior Mathematical Olympiad), which started in 2010 and follows the same format as the USAMO. A few cities in China had participated in the competition and the results were encouraging.

Similarly as in the former Soviet Union, the Mathematical Olympiad education was widely recognized in America. The book "How to Solve it" written by *George Polya* along with many other titles had been translated into many different languages. *George Polya* provided a whole series of general heuristics for solving problems of all kinds. His influence in the educational community in China should not be underestimated.

International Mathematical Olympiad

In 1956, the East European countries and the Soviet Union took the initiative to organize the IMO formally. The first International Mathematical Olympiad (IMO) was held in Brasov, Romania, in 1959. At the time, there were only seven participating countries, namely, Romania, Bulgaria, Poland, Hungary, Czechoslovakia, East Germany and the Soviet Union. Subsequently, the United States of America, United Kingdom, France,

Germany and also other countries including those from Asia joined. Today, the IMO had managed to reach almost all the developed and developing countries. Except in the year 1980 due to financial difficulties faced by the host country, Mongolia, there were already 57 Olympiads held and 109 countries and regions participating.

The mathematical topics in the IMO include Algebra, Combinatorics, Geometry, and Number theory. These areas had provided guidance for setting questions for the competitions. Other than the first few Olympiads, each IMO is normally held in mid-July every year and the test paper consists of 6 questions in all. The actual competition lasts for 2 days for a total of 9 hours where participants are required to complete 3 questions each day. Each question is 7 points which total up to 42 points. The full score for a team is 252 marks. About half of the participants will be awarded a medal, where 1/12 will be awarded a gold medal. The numbers of gold, silver and bronze medals awarded are in the ratio of 1:2:3 approximately. In the case when a participant provides a better solution than the official answer, a special award is given.

Each participating country and region will take turn to host the IMO. The cost is borne by the host country. China had successfully hosted the 31st IMO in Beijing. The event had made a great impact on the mathematical community in China. According to the rules and regulations of the IMO, all participating countries are required to send a delegation consisting of a leader, a deputy leader and 6 contestants. The problems are contributed by the participating countries and are later selected carefully by the host country for submission to the international jury set up by the host country. Eventually, only 6 problems will be accepted for use in the competition. The host country does not provide any question. The short-listed problems are subsequently translated, if necessary, in English, French, German, Spain, Russian and other working languages. After that, the team leaders will translate the problems into their own languages.

The answer scripts of each participating team will be marked by the team leader and the deputy leader. The team leader will later present the scripts of their contestants to the coordinators for assessment. If there is any dispute, the matter will be settled by the jury. The jury is formed by the various team leaders and an appointed chairman by the host country. The jury is responsible for deciding the final 6 problems for the competition. Their duties also include finalizing the grading standard, ensuring the accuracy of the translation of the problems, standardizing replies to written queries raised by participants during the competition, synchronizing differences in grading between the team leaders and the coordinators and also deciding on the cut-off points for the medals depending on the contestants' results as the difficulties of problems each year are different.

China had participated informally in the 26th IMO in 1985. Only two students were sent. Starting from 1986, except in 1998 when the IMO was held in Taiwan, China had always sent 6 official contestants to the IMO. Today, the Chinese contestants not only performed outstandingly in the IMO, but also in the International Physics, Chemistry, Informatics, and Biology Olympiads. This can be regarded as an indication that China pays great attention to the training of basic skills in mathematics and science education.

Winners of the IMO

Among all the IMO medalists, there were many of them who eventually became great mathematicians. They were also awarded the Fields Medal, Wolf Prize and Nevanlinna Prize (a prominent mathematics prize for computing and informatics). In what follows, we name some of the winners.

G. Margulis, a silver medalist of IMO in 1959, was awarded the Fields Medal in 1978. *L. Lovász*, who won the Wolf Prize in 1999, was awarded the Special Award in IMO consecutively in 1965 and 1966. *V. Drinfeld*, a gold medalist of IMO in 1969, was awarded the Fields Medal in 1990. *J.-C. Yoccoz* and *T. Gowers*, who were both awarded the Fields Medal in 1998, were gold medalists in IMO in 1974 and 1981 respectively. A silver medalist of IMO in 1985, *L. Lafforgue*, won the Fields Medal in 2002. A gold medalist of IMO in 1982, *Grigori Perelman* from Russia, was awarded the Fields Medal in 2006 for solving the final step of the Poincaré conjecture. In 1986, 1987, and 1988, *Terence Tao* won a bronze, silver, and gold medal respectively. He was the youngest participant to date in the IMO, first competing at the age of ten. He was also awarded the Fields Medal in 2006. Gold medalist of IMO 1988 and 1989, *Ngo Bau Chao*, won the Fields Medal in 2010, together with the bronze medalist of IMO 1988, *E. Lindenstrauss*. Gold medalist of IMO 1994 and 1995, *Maryam Mirzakhani* won the Fields Medal in 2014. A gold medalist of IMO in 1995, Artur Avila won the Fields Medal in 2014.

A silver medalist of IMO in 1977, *P. Shor*, was awarded the Nevanlinna Prize. A gold medalist of IMO in 1979, *A. Razborov*, was awarded the Nevanlinna Prize. Another gold medalist of IMO in 1986, *S. Smirnov*, was awarded the Clay Research Award. *V. Lafforgue*, a gold medalist of

IMO in 1990, was awarded the European Mathematical Society prize. He is *L. Lafforgue*'s younger brother.

Also, a famous mathematician in number theory, *N. Elkies*, who is also a professor at Harvard University, was awarded a gold medal of IMO in 1982. Other winners include *P. Kronheimer* awarded a silver medal in 1981 and *R. Taylor* a contestant of IMO in 1980.

Mathematical competition in China

Due to various reasons, mathematical competition in China started relatively late but is progressing vigorously.

"We are going to have our own mathematical competition too!" said *Hua Luogeng*. *Hua* is a house hold name in China. The first mathematical competition was held concurrently in Beijing, Tianjin, Shanghai and Wuhan in 1956. Due to the political situation at the time, this event was interrupted a few times. Until 1962, when the political environment started to improve, Beijing and other cities started organizing the competition though not regularly. In the era of Cultural Revolution, the whole educational system in China was in chaos. The mathematical competition came to a complete halt. In contrast, the mathematical competition in the former Soviet Union was still on-going during the war and at a time under the difficult political situation. The competitions in Moscow were interrupted only 3 times between 1942 and 1944. It was indeed commendable.

In 1978, it was the spring of science. *Hua Luogeng* conducted the Middle School Mathematical Competition for 8 provinces in China. The mathematical competition in China was then making a fresh start and embarked on a road of rapid development. *Hua* passed away in 1985. In commemorating him, a competition named *Hua Luogeng* Gold Cup was set up in 1986 for students in Grade 6 and 7 and it has a great impact.

The mathematical competitions in China before 1980 can be considered as the initial period. The problems set were within the scope of middle school textbooks. After 1980, the competitions were gradually moving towards the senior middle school level. In 1981, the Chinese Mathematical Society decided to conduct the China Mathematical Competition, a national event for high schools.

In 1981, the United States of America, the host country of IMO, issued an invitation to China to participate in the event. Only in 1985, China sent two contestants to participate informally in the IMO. The results were not encouraging. In view of this, another activity called the Winter Camp was conducted after the China Mathematical Competition. The Winter Camp was later renamed as the China Mathematical Olympiad or CMO. The winning team would be awarded the *Chern Shiing-Shen* Cup. Based on the outcome at the Winter Camp, a selection would be made to form the 6-member national team for IMO. From 1986 onwards, other than the year when IMO was organized in Taiwan, China had been sending a 6-member team to IMO. Up to 2016, China had been awarded the overall team champion for 19 times.

In 1990, China had successfully hosted the 31st IMO. It showed that the standard of mathematical competition in China has reached a level similar to other leading countries. First, the fact that China achieves the highest marks at the 31st IMO for the team is an evidence of the effectiveness of the pyramid approach in selecting the contestants in China. Secondly, the Chinese mathematicians had simplified and modified over 100 problems and submitted them to the team leaders of the 35 countries for their perusal. Eventually, 28 problems were recommended. At the end, 5 problems were chosen (IMO requires 6 problems). This is another evidence to show that China has achieved the highest quality in setting problems. Thirdly, the answer scripts of the participants were marked by the various team leaders and assessed by the coordinators who were nominated by the host countries. China had formed a group 50 mathematicians to serve as coordinators who would ensure the high accuracy and fairness in marking. The marking process was completed half a day earlier than it was scheduled. Fourthly, that was the first ever IMO organized in Asia. The outstanding performance by China had encouraged the other developing countries, especially those in Asia. The organizing and coordinating work of the IMO by the host country was also reasonably good.

In China, the outstanding performance in mathematical competition is a result of many contributions from the all quarters of mathematical community. There are the older generation of mathematicians, middle-aged mathematicians and also the middle and elementary school teachers. There is one person who deserves a special mention and he is *Hua Luogeng*. He initiated and promoted the mathematical competition. He is also the author of the following books: Beyond *Yang hui*'s Triangle, Beyond the *pi* of *Zu Chongzhi*, Beyond the Magic Computation of *Sun-zi*, Mathematical Induction, and Mathematical Problems of Bee Hive. These were his books derived from mathematics competitions. When China resumed mathematical competition in 1978, he participated in setting problems and giving critique to solutions of the problems. Other outstanding books derived from the Chinese mathematics competitions are: Symmetry by *Duan Xuefu*, Lattice and Area by *Min Sihe*, One Stroke Drawing and Postman Problem by *Jiang Boju*.

After 1980, the younger mathematicians in China had taken over from the older generation of mathematicians in running the mathematical competition. They worked and strived hard to bring the level of mathematical competition in China to a new height. *Qiu Zonghu* is one such outstanding representative. From the training of contestants and leading the team 3 times to IMO to the organizing of the 31th IMO in China, he had contributed prominently and was awarded the *P. Erdős* prize.

Preparation for IMO

Currently, the selection process of participants for IMO in China is as follows.

First, the China Mathematical Competition, a national competition for high Schools, is organized on the second Sunday in September every year. The objectives are: to increase the interest of students in learning mathematics, to promote the development of co-curricular activities in mathematics, to help improve the teaching of mathematics in high schools, to discover and cultivate the talents and also to prepare for the IMO. This has happened since 1981. Currently there are about 500,000 participants taking part.

Through the China Mathematical Competition, around 350 of students are selected to take part in the China Mathematical Olympiad or CMO, that is, the Winter Camp. The CMO lasts for 5 days and is held in December every year. The types and difficulties of the problems in CMO are very much similar to the IMO. There are also 3 problems to be completed within 4.5 hours each day. However, the score for each problem is 21 marks which add up to 126 marks in total. Starting from 1990, the Winter Camp instituted the *Chern Shiing-Shen* Cup for team championship. In 1991, the Winter Camp was officially renamed as the China Mathematical Olympiad (CMO). It is similar to the highest national mathematical competition in the former Soviet Union and the United States.

The CMO awards the first, second and third prizes. Among the participants of CMO, about 60 students are selected to participate in the training for IMO. The training takes place in March every year. After 6 to 8 tests and another 2 rounds of qualifying examinations, only 6 contestants are short-listed to form the China IMO national team to take part in the IMO in July.

Besides the China Mathematical Competition (for high schools), the Junior Middle School Mathematical Competition is also developing well. Starting from 1984, the competition is organized in April every year by the Popularization Committee of the Chinese Mathematical Society. The various provinces, cities and autonomous regions would rotate to host the event. Another mathematical competition for the junior middle schools is also conducted in April every year by the Middle School Mathematics Education Society of the Chinese Educational Society since 1998 till now.

The *Hua Luogeng* Gold Cup, a competition by invitation, had also been successfully conducted since 1986. The participating students comprise elementary six and junior middle one students. The format of the competition consists of a preliminary round, semi-finals in various provinces, cities and autonomous regions, then the finals.

Mathematical competition in China provides a platform for students to showcase their talents in mathematics. It encourages learning of mathematics among students. It helps identify talented students and to provide them with differentiated learning opportunity. It develops co-curricular activities in mathematics. Finally, it brings about changes in the teaching of mathematics.

Contents

China Mathematical Competition

(Jilin)

Commissioned by the Chinese Mathematical Society (CMS), the 2014 China Mathematical Competition was organized by Jilin Mathematical Society. The CMS set the problems; and the commissioned province was responsible for the other organizational issues.

The contest was held on Sunday, September 14th, 2014. A noticeable change was in the date of the contest — after the modification of the entrance exam exemption policy, the contest date was moved roughly one month earlier than previous years, so most of the students could have proper preparation schedules for the national college entrance examination.

1366 students from 31 regions got the regional first prize. 346 students were qualified for the 30th China Mathematical Olympiad, to be held in Chongqing.

Part I Short-Answer Questions (Questions 1-8, 8 marks each)

1 If two positive numbers a and b satisfy $2 + \log_2 a = 3 + \log_3 b = \log_6(a + b)$, then the value of $\dfrac{1}{a} + \dfrac{1}{b}$ is _____.

Solution Set $2 + \log_2 a = 3 + \log_3 b = \log_6(a + b) = k$, then $a = 2^{k-2}$, $b = 3^{k-3}$, and $a + b = 6^k$; therefore

$$\frac{1}{a} + \frac{1}{b} = \frac{a+b}{ab} = \frac{6^k}{2^{k-2} \times 3^{k-3}} = 2^2 \times 3^3 = 108. \qquad \square$$

2 Let M and m be the maximum and minimum elements of the set

$$\left\{ \frac{3}{a} + b : 1 \le a \le b \le 2 \right\},$$

respectively, then the value of $M - m$ is _____.

Solution From $1 \le a \le b \le 2$ we get $\frac{3}{a} + b \le \frac{3}{1} + 2 = 5$, the maximum value $M = 5$ is achieved when $a = 1$ and $b = 2$.

On the other hand,

$$\frac{3}{a} + b \ge \frac{3}{a} + a \ge 2\sqrt{\frac{3}{a} \cdot a} = 2\sqrt{3}.$$

And the minimum value $m = 2\sqrt{3}$ is achieved when $a = b = \sqrt{3}$.

So, $M - m = 5 - 2\sqrt{3}$.   □

3 The function $f(x) = x^2 + a|x - 1|$ is monotone increasing on $[0, +\infty)$, then the range of the real number a is _____.

Solution On $[1, +\infty)$, the function $f(x) = x^2 + ax - a$ is monotone increasing, an equivalent condition is that $-\frac{a}{2} \le 1$, i.e. $a \ge -2$.

On $[0, 1]$, the function $f(x) = x^2 - ax + a$ is monotone increasing, which is equivalent to the condition that $\frac{a}{2} \le 0$, i.e. $a \le 0$.

So, the range of a is $[-2, 0]$.   □

4 The sequence $\{a_n\}$ satisfies $a_1 = 2$, $a_{n+1} = \frac{2(n+2)}{n+1} a_n$ $(n \in \mathbb{N}^*)$, then

$$\frac{a_{2014}}{a_1 + a_2 + \cdots + a_{2013}} = \underline{\qquad}.$$

Solution We have

$$a_n = \frac{2(n+1)}{n} a_{n-1} = \frac{2(n+1)}{n} \frac{2n}{n-1} a_{n-2} = \cdots$$

$$= \frac{2(n+1)}{n} \frac{2n}{n-1} \cdots \frac{2 \cdot 3}{2} a_1 = 2^{n-1}(n+1).$$

Denote S_n the sum of the first n terms in the sequence $\{a_n\}$, so

$$S_n = 2 + 2 \times 3 + 2^2 \times 4 + \cdots + 2^{n-1}(n+1),$$

therefore

$$2S_n = 2 \times 2 + 2^2 \times 3 + 2^3 \times 4 + \cdots + 2^n(n+1).$$

Take the difference of the above two, we get

$$S_n = 2^n(n+1) - (2^{n-1} + 2^{n-2} + \cdots + 2 + 2)$$

$$= 2^n(n+1) - 2^n = 2^n n.$$

So,

$$\frac{a_{2014}}{a_1 + a_2 + \cdots + a_{2013}} = \frac{2^{2013} \times 2015}{2^{2013} \times 2013} = \frac{2015}{2013}. \qquad \square$$

5 In a square pyramid $P - ABCD$, each lateral face is an equilateral triangle with side length 1, M and N are the midpoints of AB and BC, respectively, then the distance between the skew lines MN and PC is _____.

Solution On the base face, let O be the intersection of the diagonals AC and BD, make the perpendicular to the line MN through C and meet MN at H.

Since PO is perpendicular to the base face, so $PO \perp CH$. Also we have $AC \perp CH$, so CH is perpendicular to the plane POC, and $CH \perp PC$.

So the segment CH is perpendicular to both MN and PC, and its length is

$$CH = \frac{\sqrt{2}}{2}CN = \frac{\sqrt{2}}{4}.$$

So, the distance between MN and PC is $\dfrac{\sqrt{2}}{4}$. $\qquad \square$

6 F_1 and F_2 are the foci of the ellipse Γ. A line through F_1 meets Γ at P and Q. If $|PF_2| = |F_1F_2|$ and $3|PF_1| = 4|QF_1|$, the ratio of the semi-minor axis of Γ to its semi-major axis is _____.

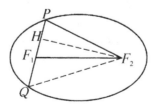

Fig. 6.1

Solution Without loss of generality, assume $|PF_1| = 4$ and $|QF_1| = 3$. Denote a and b the semi-major axis and semi-minor axis, respectively, and let $2c$ be the distance between the foci. So $|PF_1| = |F_1F_2| = 2c$, and by the property of the ellipse,

$$2a = |QF_1| + |QF_2| = |PF_1| + |PF_2| = 2c + 4.$$

Therefore,

$$|QF_2| = |PF_1| + |PF_2| - |QF_1| = 2c + 1.$$

Let H be the midpoint of PF_1, then $|F_1H| = 2$ and $|QH| = 5$, and $F_2H \perp PF_1$. By Pythagorean theorem,

$$|QF_2|^2 - |QH|^2 = |F_2H|^2 = |F_1F_2|^2 - |F_1H|^2,$$

so, $(2c + 1)^2 - 5^2 = (2c)^2 - 2^2$. We get $c = 5$, then $a = 7$ and $b = 2\sqrt{6}$. The ratio of the semi-minor axis of Γ to its semi-major axis is $b : a = \dfrac{2\sqrt{6}}{7}$.

$\square$

7 The incircle of an equilateral triangle ABC has radius 2 and has the point I as its center. A point P satisfies $|PI| = 1$, then the maximum ratio of the area of $\triangle APB$ to the area of $\triangle APC$ is _____.

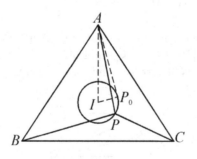

Fig. 7.1

Solution The trace of P is the unit circle centered at I, call this circle K.

Assume $\angle BAP = \alpha$. Take a point P_0 on K so α achieves its maximum value α_0. P_0 is inside the $\angle IAC$, and AP_0 is tangent to the circle K at P_0.

Since $0 < \alpha \le \alpha_0 < \pi/3$,

$$\frac{S_{\triangle APB}}{S_{\triangle APC}} = \frac{\frac{1}{2} AP \cdot AB \cdot \sin \alpha}{\frac{1}{2} AP \cdot AC \cdot \sin \left(\frac{\pi}{3} - \alpha \right)} = \frac{\sin \alpha}{\sin \left(\frac{\pi}{3} - \alpha \right)} \qquad \text{①}$$

$$\le \frac{\sin \alpha_0}{\sin \left(\frac{\pi}{3} - \alpha_0 \right)} = \frac{\sin \left(\frac{\pi}{6} + \theta \right)}{\sin \left(\frac{\pi}{6} - \theta \right)},$$

where $\theta = \alpha_0 - \pi/6 = \angle I A P_0$.

Since $\angle A P_0 I = \frac{\pi}{2}$,

$$\sin \theta = \frac{IP_0}{AI} = \frac{1}{2r} = \frac{1}{4}.$$

Therefore, $\cot \theta = \sqrt{15}$, and

$$\frac{\sin \left(\frac{\pi}{6} + \theta \right)}{\sin \left(\frac{\pi}{6} - \theta \right)} = \frac{\frac{1}{2} \cos \theta + \frac{\sqrt{3}}{2} \sin \theta}{\frac{1}{2} \cos \theta - \frac{\sqrt{3}}{2} \sin \theta} = \frac{\cot \theta + \sqrt{3}}{\cot \theta - \sqrt{3}} \qquad \text{②}$$

$$= \frac{\sqrt{15} + \sqrt{3}}{\sqrt{15} - \sqrt{3}} = \frac{3 + \sqrt{5}}{2}.$$

By ① and ②, when $P = P_0$, $S_{\triangle APB} : S_{\triangle APC}$ achieves its maximum value $\dfrac{3 + \sqrt{5}}{2}$. □

8 A, B, C, D are four points in the space that are not coplanar. For any pair of two distinct points, connect the segment between them with probability $1/2$, independently. The probability that A and B are connected through a polygonal chain consists of one or more segments is _____.

Solution There are 2 situations about the direct segment between each pair of points, there are $2^6 = 64$ combined outcomes. We count the number of outcomes where A and B can be connected by a polygonal chain.

(1) The segment AB is connected. There are $2^5 = 32$ such outcomes.

(2) The segment AB is not connected, but CD is connected. In this case, A and B can be connected by a polygonal chain if and only if both A and B are connected to at least one of C and D. There are $(2^2 - 1) \times (2^2 - 1) = 9$ such outcomes.

(3) Both segments AB and CD are not connected. In this case at least one of C and D should be connected to both A and B. By inclusion-exclusion, the number of such outcomes is $2^2 + 2^2 - 1 = 7$.

There are $32 + 9 + 7 = 48$ outcomes counted in the above cases. The probability in question is $\dfrac{48}{64} = \dfrac{3}{4}$. □

Part II Word Problems (Questions 9-11, 56 marks in total)

9 (16 marks) In the Cartesian coordinate plane xOy, P is a moving point not on the x-axis that satisfies: There are two tangent lines through P to the parabola $y^2 = 4x$, and the line ℓ_P passing through the two points of tangency is perpendicular to the line PO.
The line ℓ_P insects PO and the x-axis at Q and R, respectively.

(1) Prove that R is a fixed point.

(2) Find the minimum value of $\dfrac{|PQ|}{|QR|}$.

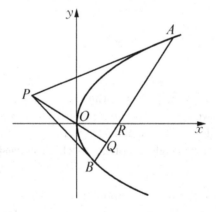

Fig. 9.1

Solution (1) Suppose $P(a, b)$, where $b \neq 0$; it is obvious that $a \neq 0$ as well. Suppose the two points of tangency are $A(x_1, y_1)$ and $B(x_2, y_2)$, then the equations for the lines PA and PB are

$$yy_1 = 2(x + x_1), \qquad\qquad ①$$

and

$$yy_2 = 2(x + x_2).$$ ②

The coordinates of P satisfies both ① and ②, so (x_1, y_1) and (x_2, y_2) — the coordinates of A and B — both satisfy the equation

$$by = 2(x + a).$$ ③

Therefore ③ is the equation for the line AB.

The slopes of the lines PO and AB are $\dfrac{b}{a}$ and $\dfrac{2}{b}$, respectively. Since $PO \perp AB$, we have $\dfrac{b}{a} \cdot \dfrac{2}{b} = 1$, so $a = -2$.

Now ③ becomes $y = \dfrac{2}{b}(x - 2)$. The intersection of the line AB and the x-axis is the fixed point $R(2, 0)$.

(2) We had $a = -2$. The slope of the line PO is $k_1 = -\dfrac{b}{2}$, and the slope of the line PR is $k_2 = -\dfrac{b}{4}$. Denote $\alpha = \angle OPR$, we have $\alpha < \pi/2$, and

$$\frac{|PQ|}{|QR|} = \frac{1}{\tan a} = \left| \frac{1 + k_1 k_2}{k_1 - k_2} \right| = \left| \frac{1 + \left(-\dfrac{b}{2}\right)\left(-\dfrac{b}{4}\right)}{-\dfrac{b}{2} + \dfrac{b}{4}} \right|$$

$$= \frac{8 + b^2}{2|b|} \geq \frac{2\sqrt{8b^2}}{2|b|} = 2\sqrt{2}.$$

When $b = \pm 2\sqrt{2}$, $\dfrac{|PQ|}{|QR|}$ achieves its minimum value $2\sqrt{2}$. $\qquad\square$

10 (20 marks) The sequence $\{a_n\}$ satisfies $a_1 = \pi/6$, and $a_{n+1} = \arctan(\sec a_n)$ for all $n \in \mathbb{N}^*$. Find the positive integer m for which

$$\sin a_1 \cdot \sin a_2 \cdot \ldots \cdot \sin a_m = \frac{1}{100}.$$

Solution From the definition of $\{a_n\}$, we see that $a_{n+1} \in (-\pi/2, \pi/2)$ for any positive integer n, and

$$\tan a_{n+1} = \sec a_n.$$ ①

Since $\sec a_n > 0$, so $a_{n+1} \in (0, \pi/2)$. From ①, we get $\tan^2 a_{n+1} = \sec^2 a_n = 1 + \tan^2 a_n$. So,

$$\tan^2 a_n = n - 1 + \tan^2 a_1 = n - 1 + \frac{1}{3} = \frac{3n - 2}{3},$$

i.e.,

$$\tan a_n = \sqrt{\frac{3n-2}{3}}.$$

Now

$$\sin a_1 \cdot \sin a_2 \cdot \ldots \cdot \sin a_m = \frac{1}{100} = \frac{\tan a_1}{\sec a_1} \cdot \frac{\tan a_2}{\sec a_2} \cdot \ldots \cdot \frac{\tan a_m}{\sec a_m}$$

$$= \frac{\tan a_1}{\tan a_2} \cdot \frac{\tan a_2}{\tan a_3} \cdot \ldots \cdot \frac{\tan a_m}{\tan a_{m+1}} \quad (\text{by } \textcircled{1})$$

$$= \frac{\tan a_1}{\tan a_{m+1}} = \sqrt{\frac{1}{3m+1}}.$$

We want $\sqrt{\dfrac{1}{3m+1}} = \dfrac{1}{100}$, so $m = 3333$. $\square$

11 (20 marks) Determine all the complex numbers α, so that for any complex numbers z_1 and z_2, where $|z_1|, |z_2| < 1$ and $z_1 \neq z_2$, we have

$$(z_1 + \alpha)^2 + \alpha \overline{z_1} \neq (z_2 + \alpha)^2 + \alpha \overline{z_2}.$$

Solution   Let $f_\alpha(z) = (z + \alpha)^2 + \alpha \overline{z}$.

$$f_\alpha(z_1) - f_\alpha(z_2) = (z_1 + \alpha)^2 + \alpha \overline{z_1} - (z_2 + \alpha)^2 - \alpha \overline{z_2}$$

$$= (z_1 + z_2 + 2\alpha)(z_1 - z_2) + \alpha(\overline{z_1} - \overline{z_2}). \qquad \textcircled{1}$$

Suppose there are complex numbers $z_1 \neq z_2$ such that $|z_1|, |z_2| < 1$ and $f_\alpha(z_1) = f_\alpha(z_2)$. By $\textcircled{1}$,

$$|\alpha(\overline{z_1} - \overline{z_2})| = |-(z_1 + z_2 + 2\alpha)(z_1 - z_2)|.$$

Note that $|\overline{z_1} - \overline{z_2}| = |z_1 - z_2|$, so

$$|\alpha| = |z_1 + z_2 + 2\alpha| \geq 2|\alpha| - |z_1| - |z_2| > 2|\alpha| - 2;$$

i.e., $|\alpha| < 2$.

On the other hand, whenever α is a complex with $|\alpha| < 2$, set $z_1 = -\alpha/2 + i\beta$ and $z_2 = -\alpha/2 - i\beta$, where β is any real in $(0, 1 - |\alpha|/2)$. Then $z_1 \neq z_2$, and

$$\left| -\frac{\alpha}{2} \pm i\beta \right| \leq \left| -\frac{\alpha}{2} \right| + |\beta| < 1.$$

So $|z_1|, |z_2| < 1$. Now put

$$z_1 + z_2 = -\alpha, \quad z_1 - z_2 = 2i\beta, \quad \overline{z_1} - \overline{z_2} = \overline{2i\beta} = -2i\beta.$$

in $\boxed{1}$, we get

$$f_\alpha(z_1) - f_\alpha(z_2) = \alpha \cdot 2i\beta + \alpha \cdot (-2i\beta) = 0;$$

so, $f_\alpha(z_1) = f_\alpha(z_2)$.

We conclude that the set of complexes in question is $\{\alpha \in \mathbb{C} : |\alpha| \geq 2\}$.

$\square$

China Mathematical Competition

(Sichuan)

Commissioned by the Chinese Mathematical Society (CMS), the 2015 China Mathematical Competition was organized by the Sichuan Mathematical Society. The contest was held on Sunday, September 13th, 2015.

1408 students from 31 regions got the regional first prize. 350 students were qualified for the 31st China Mathematical Olympiad, to be held in Yingtan, Jiangxi.

Part I Short-Answer Questions (Questions 1-8, 8 marks each)

1 Suppose a and b are two distinct reals, and the quadratic function $f(x) = x^2 + ax + b$ satisfies $f(a) = f(b)$, then the value of $f(2)$ is _____.

Solution By the condition and the symmetry of the graph for the quadratic function, we have $\dfrac{a+b}{2} = -\dfrac{a}{2}$, i.e., $2a + b = 0$. So

$$f(2) = 4 + 2a + b = 4. \qquad \square$$

2 α is a real number such that $\cos \alpha = \tan \alpha$, then the value of $\dfrac{1}{\sin \alpha} + \cos^4 \alpha$ is _____.

Solution From the condition, $\cos^2 \alpha = \sin \alpha$. Repeatedly using this and noting that $\cos^2 \alpha + \sin^2 \alpha = 1$, we get

$$\frac{1}{\sin \alpha} + \cos^4 \alpha = \frac{\cos^2 \alpha + \sin^2 \alpha}{\sin \alpha} + \sin^2 \alpha$$

$$= (1 + \sin \alpha) + (1 - \cos^2 \alpha)$$

$$= 2 + \sin \alpha - \cos^2 \alpha = 2. \qquad \square$$

3 A series $\{z_n\}$ of complex numbers satisfies $z_1 = 1$ and, for each $n = 1, 2, \ldots,$ $z_{n+1} = \overline{z_n} + 1 + ni$, where i is the imaginary unit, and $\overline{z_n}$ is the conjugate of z_n. The value of z_{2015} is _____.

Solution From the condition, for any positive integer n, we have

$$z_{n+2} = \overline{z_{n+1}} + 1 + (n+1)i$$

$$= \overline{\overline{z_n} + 1 + ni} + 1 + (n+1)i$$

$$= z_n + 2 + i.$$

So $z_{2015} = z_1 + 1007 \times (2 + i) = 2015 + 1007i.$ $\qquad \square$

4 In a rectangle $ABCD$, $AB = 2$ and $AD = 1$. A point P moving on the edge DC (including the end points) and a point Q moving on the extension of the edge CB (including B) satisfy $|\overrightarrow{DP}| = |\overrightarrow{BQ}|$. The minimum value of the dot product $\overrightarrow{PA} \cdot \overrightarrow{PQ}$ is _____.

Solution Without loss of generality, denote $A(0,0)$, $B(2,0)$, $D(0,1)$. Denote the coordinates of P by $(t, 1)$, where $0 \le t \le 2$; by $|\overrightarrow{DP}| = |\overrightarrow{BQ}|$, the coordinates of Q is $(2, -t)$. So $\overrightarrow{PA} = (-t, -1)$ and $\overrightarrow{PQ} = (2-t, -t-1)$, and

$$\overrightarrow{PA} \cdot \overrightarrow{PQ} = (-t)(2-t) + (-1)(-t-1)$$

$$= t^2 - t - 1$$

$$= \left(t - \frac{1}{2}\right)^2 + \frac{3}{4}$$

$$\ge \frac{3}{4}.$$

When $t = \frac{1}{2}$, $\overrightarrow{PA} \cdot \overrightarrow{PQ}$ achieves the minimum $\frac{3}{4}$. $\qquad \square$

5 Uniformly randomly pick 3 edges from the edges of a unit cube, the probability that they are pairwise non-coplanar is _____.

Solution Denote the cube by $ABCD - EFGH$. There are $\binom{12}{3} = 220$ ways to pick 3 edges out of its 12 edges.

Now we count the number of ways where the 3 picked edges are pairwise non-coplanar. The edges of the cube have 3 different pairwise non-parallel directions, namely, AB, AD, and AE. Parallel edges are coplanar, so the 3 edges we pick must have one from each of the directions. We first pick an edge parallel to AB, there are 4 choices. Without loss of generality, suppose we picked AB. Next we pick an edge parallel to AD, we have only 2 choices — EH or FG. If we picked EH in the second step, there is exactly one choice, CG, for the edge parallel to AE; similarly, if we picked FG in the second step, there is exactly one choice, DH, for the edge parallel to AE.

So, the number of ways to pick 3 edges that are pairwise non-coplanar is $4 \times 2 = 8$, and the probability in question is $\dfrac{8}{220} = \dfrac{2}{55}$. □

6 In the Cartesian coordinate system xOy, the area of the region corresponding to the point set $K = \{(x, y) : (|x| + |3y| - 6)(|3x| + |y| - 6) \le 0\}$ is _____.

Solution Let $K_1 = \{(x, y) : |x| + |3y| - 6 \le 0\}$. First consider the graph of K_1 in quadrant I. Here we have $x + 3y \le 6$, so they are the points inside $\triangle OCD$ as in Figure 6.1. By symmetry, the region corresponding to K_1 is the diamond $ABCD$ together with its interior.

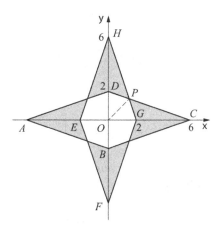

Fig. 6.1

Similarly, the set $K_2 = \{(x, y) : |3x| + |y| - 6 \leq 0\}$ corresponds to the region consists of the diamond $EFGH$ and its interior.

By the definition of K, the region corresponding to K, except for the boundary, consists of those points covered by exactly one of K_1 and K_2. So its area is the area S of the shaded region in the figure.

The equation for the line CD is $x + 3y = 6$, and the equation for the line GH is $3x + y = 6$; it is easy to get that their intersection is $P(3/2, 3/2)$. By symmetry,

$$S = 8S_{\triangle CPG} = 8 \times \frac{1}{2} \times 4 \times \frac{3}{2} = 24. \qquad \square$$

7 ω is a positive real. Suppose there are two reals a and b such that $\pi \leq a < b \leq 2\pi$ and $\sin \omega a + \sin \omega b = 2$, then the range of ω is

_____.

Solution $\sin \omega a + \sin \omega b = 2$ implies $\sin \omega a = \sin \omega b = 1$. When $a < b$ range over $[\pi, 2\pi]$, ωa and ωb range over any pair of two points in $[\omega \pi, 2\omega \pi]$. So, the condition is equivalent to: there are integers $k < l$ such that

$$\omega \pi \leq 2k\pi + \frac{\pi}{2} < 2l\pi + \frac{\pi}{2} \leq 2\omega\pi. \qquad ①$$

When $\omega \geq 4$, the length of the interval $[\omega\pi, 2\omega\pi]$ is no less than 4π, clearly there are $k < l$ satisfying ①.

When $0 < \omega < 4$, note that $[\omega\pi, 2\omega\pi] \subseteq (0, 8\pi)$, we consider the following cases.

(1) $\omega\pi \leq \dfrac{\pi}{2} < \dfrac{5\pi}{2} \leq 2\omega\pi$ implies $\omega \leq \dfrac{1}{2}$ and $\omega \geq \dfrac{5}{4}$, which is impossible.

(2) $\omega\pi \leq \dfrac{5\pi}{2} < \dfrac{9\pi}{2} \leq 2\omega\pi$ implies $\dfrac{9}{4} \leq \omega \leq \dfrac{5}{2}$.

(3) $\omega\pi \leq \dfrac{9\pi}{2} < \dfrac{13\pi}{2} \leq 2\omega\pi$ implies $\dfrac{13}{4} \leq \omega \leq \dfrac{9}{2}$. Since we assumed here $\omega < 4$, so $\dfrac{13}{4} \leq \omega < 4$.

In conclusion, we have the range of ω as $\left[\dfrac{9}{4}, \dfrac{5}{2}\right] \cup \left[\dfrac{13}{4}, +\infty\right)$. $\qquad \square$

8 Consider any 4-digit decimal numbers $\overline{abcd}$, where $1 \leq a \leq 9$, $0 \leq b, c, d \leq 9$. Call it *type P* if $a > b$, $b < c$, and $c > d$; call it *type Q* if $a < b$, $b > c$, and $c < d$. Denote by $N(P)$ and $N(Q)$ the number of type P and type Q integers, respectively. The value of $N(P) - N(Q)$ is _____.

Solution Denote the set of type P and type Q numbers by A and B, respectively. Let A_0 be the set of numbers in A with $d = 0$, and A_1 be the set of numbers in A with $d \neq 0$.

For any element $\overline{abcd} \in A_1$, we reverse its digits, and it is easy to see that $\overline{dcba} \in B$. Conversely, every $\overline{dcba} \in B$ corresponds to a unique $\overline{abcd} \in A_1$. This is a one-one correspondence between A_1 and B, so

$$N(P) - N(Q) = |A| - |B| = |A_0| + |A_1| - |B| = |A_0|.$$

Now we compute $|A_0|$. For any $\overline{abc0} \in A_0$, b can be any of $0, 1, \ldots, 9$, and for each b, by $b < a \leq 9$ and $b < c \leq 9$, each of a and c has $9 - b$ choices. Thus

$$|A_0| = \sum_{b=0}^{9}(9-b)^2 = \sum_{k=1}^{9} k^2 = \frac{9 \times 10 \times 19}{2} = 285.$$

Hence $N(P) - N(Q) = 285.$ □

Part II Word Problems (Questions 9-11, 56 marks in total)

9 (16 marks) Real numbers a, b, and c satisfy $2^a + 4^b = 2^c$ and $4^a + 2^b = 4^c$, determine the minimum value of c.

Solution Denote 2^a, 2^b, and 2^c by x, y, and z, respectively. We have $x, y, z > 0$, $x + y^2 = z$, and $x^2 + y = z^2$. So

$$z^2 - y = x^2 = (z - y^2)^2 = z^2 - 2y^2 z + y^4.$$

Together with the AM-GM inequality, we have

$$z = \frac{y^4 + y}{2y^2} = \frac{1}{4}\left(2y^2 + \frac{1}{y} + \frac{1}{y}\right)$$

$$\geq \frac{1}{4} \cdot 3 \sqrt[3]{2y^2 \cdot \frac{1}{y} \cdot \frac{1}{y}} = \frac{3}{4}\sqrt[3]{2}.$$

When $2y^2 = \frac{1}{y}$, i.e., $y = \frac{1}{\sqrt[3]{2}}$, z achieves its minimum $\frac{3}{4}\sqrt[3]{2}$ and the corresponding $x = \frac{\sqrt[3]{2}}{4}$.

Since $c = \log_2 z$, so the minimum of c is $\log_2\left(\frac{3}{4}\sqrt[3]{2}\right) = \log_2 3 - \frac{5}{3}$. □

10 (20 marks) a, b, c, and d are 4 rationals such that

$$\{a_i a_j : 1 \leq i < j \leq 4\} = \left\{-24, -2, -\frac{3}{2}, -\frac{1}{8}, 1, 3\right\}.$$

Determine the value of $a_1 + a_2 + a_3 + a_4$.

Solution   Observe that $a_i a_j$, $1 \leq i < j \leq 4$, are 6 distinct numbers, and no two of them are opposite numbers, so their absolute values are distinct. Without loss of generality, suppose $|a_1| < |a_2| < |a_3| < |a_4|$, then among the $|a_i a_j|$'s, $1 \leq i < j \leq 4$, the smallest is $|a_1 a_2|$, the second smallest is $|a_1 a_3|$, the largest is $|a_3 a_4|$, and the second largest is $|a_2 a_4|$. So we have

$$\begin{cases} a_1 a_2 = -\dfrac{1}{8}, \\[2mm] a_1 a_3 = 1, \\[2mm] a_2 a_4 = 3, \\[2mm] a_3 a_4 = -24. \end{cases}$$

These imply

$$a_2 = -\frac{1}{8a_1}, \quad a_3 = \frac{1}{a_1}, \quad a_4 = \frac{3}{a_2} = -24a_1;$$

so

$$\{a_2 a_3, a_1 a_4\} = \left\{-\frac{1}{8a_1^2}, -24a_1^2\right\} = \left\{-2, -\frac{3}{2}\right\}.$$

Note that $a_1 \in \mathbb{Q}$, we get $a_1 = \pm\frac{1}{4}$.

It is easy to verify that $a_1 = \frac{1}{4}$, $a_2 = -\frac{1}{2}$, $a_3 = 4$, and $a_4 = -6$ satisfy the condition in the problem statement; so do $a_1 = -\frac{1}{4}$, $a_2 = \frac{1}{2}$, $a_3 = -4$, and $a_4 = 6$.

So $a_1 + a_2 + a_3 + a_4 = \pm\frac{9}{4}$.   □

11 (20 marks) In the Cartesian coordinate system xOy, F_1 and F_2 are the left and right foci of the ellipse $\dfrac{x^2}{2} + y^2 = 1$, respectively. A line ℓ, which does not pass through F_1, intersects the ellipse at two points A and B; the distance from F_2 to ℓ is d; and the slopes of the lines AF_1, ℓ, and BF_1 form an arithmetic progression. Determine the range of d.

Solution By the equation of the ellipse, the coordinates of F_1 and F_2 are $(-1, 0)$ and $(1, 0)$, respectively.

Suppose the equation for ℓ is $y = kx + m$, and the coordinates for A and B are $A(x_1, y_1)$ and $B(x_2, y_2)$, so x_1 and x_2 satisfy the equation $\dfrac{x^2}{2} + (kx + m)^2 = 1$, i.e.,

$$(2k^2 + 1)x^2 + 4kmx + (2m^2 - 2) = 0. \qquad \textcircled{1}$$

Since A and B are two distinct points, and the slope of ℓ exists, x_1 and x_2 must be two different roots of $\textcircled{1}$, therefore the discriminant of $\textcircled{1}$

$$\Delta = (4km)^2 - 4(2k^2 + 1)(2m^2 - 2) = 8(2k^2 + 1 - m^2) > 0,$$

i.e.,

$$2k^2 + 1 > m^2. \qquad \textcircled{2}$$

Now, $\dfrac{y_1}{x_1 + 1}$, k, and $\dfrac{y_2}{x_2 + 1}$ are the slopes of AF_1, ℓ, and BF_1, respectively, and they form an arithmetic progression, so $\dfrac{y_1}{x_1 + 1} + \dfrac{y_2}{x_2 + 1} = 2k$, note that $y_1 = kx_1 + m$ and $y_2 = kx_2 + m$, we get

$$(kx_1 + m)(x_2 + 1) + (kx_2 + m)(x_1 + 1) = 2k(x_1 + 1)(x_2 + 1).$$

After simplification, we get

$$(m - k)(x_1 + x_2 + 2) = 0.$$

When $m = k$, the equation for ℓ is $y = kx + k$, and ℓ passed through F_1, which violates our condition.

So $x_1 + x_2 + 2 = 0$. By $\textcircled{1}$ and Vieta's formulas,

$$\frac{4km}{2k^2 + 1} = -(x_1 + x_2) = 2,$$

i.e.

$$m = k + \frac{1}{2k}. \qquad \textcircled{3}$$

By $\textcircled{2}$ and $\textcircled{3}$, $2k^2 + 1 > m^2 = \left(k + \dfrac{1}{2k}\right)^2$. Simplify and we get $k^2 > \dfrac{1}{4k^2}$, which is equivalent to $|k| > \dfrac{\sqrt{2}}{2}$.

Conversely, when m and k satisfy ③ and $|k| > \dfrac{\sqrt{2}}{2}$, ℓ does not pass through F_1 — otherwise $m = k$ contradicts ③. We deduce in turn that m and k also satisfy ②, so ℓ intersects the ellipse at two different points A and B, meanwhile the existence of the slopes of AF_1, ℓ, and BF_1 are assured — otherwise one of x_1 and x_2 is -1, together with $x_1 + x_2 + 2 = 0$, we have $x_1 = x_2 = -1$, which is impossible here since ② means ① has two distinct real roots.

The distance from $F_2(1, 0)$ to $\ell : y = kx + m$ is

$$d = \frac{|k + m|}{\sqrt{1 + k^2}} = \frac{1}{\sqrt{1 + k^2}} \left| 2k + \frac{1}{2k} \right| = \frac{1}{\sqrt{\dfrac{1}{k^2} + 1}} \left(2 + \frac{1}{2k^2} \right).$$

Note that $|k| > \dfrac{\sqrt{2}}{2}$, let $t = \sqrt{\dfrac{1}{k^2} + 1}$, then t ranges over $(1, \sqrt{3})$, and the above can be written as

$$d = \frac{1}{t} \left(\frac{t^2}{2} + \frac{3}{2} \right) = \frac{1}{2} \left(t + \frac{3}{t} \right). \qquad \text{④}$$

The function $f = \dfrac{1}{2} \left(t + \dfrac{3}{t} \right)$ is monotone decreasing on $[1, \sqrt{3}]$. By ④, we have $f(\sqrt{3}) < d < f(1)$, so $d \in (\sqrt{3}, 2)$. $\qquad \square$

China Mathematical Competition (Complementary Test)

1 (40 marks) Real numbers a, b, c satisfy $a + b + c = 1$ and $abc > 0$. Prove that

$$ab + bc + ca < \frac{\sqrt{abc}}{2} + \frac{1}{4}.$$

Proof (First proof). When $ab + bc + ca \le \frac{1}{4}$, the inequality obviously holds.

Suppose $ab + bc + ca > \frac{1}{4}$, and we may assume $a = \max\{a, b, c\}$. Since $a + b + c = 1$, $a \ge \frac{1}{3}$. We have

$$ab + bc + ca - \frac{1}{4} \le \frac{(a+b+c)^2}{3} - \frac{1}{4} = \frac{1}{12} \le \frac{a}{4}, \qquad ①$$

and

$$ab + bc + ca - \frac{1}{4} = a(b + c) - \frac{1}{4} + bc = a(1 - a) - \frac{1}{4} + bc$$

$$\le \frac{1}{4} - \frac{1}{4} + bc = bc. \qquad ②$$

The equality holds in ① when $a = 1/3$; and the equality holds in ② when $a = 1/2$; so the equality does not hold in ① and ②

19

simultaneously. Since $ab + bc + ca - \dfrac{1}{4} > 0$. Multiply ① and ② we get

$$\left(ab + bc + ca - \frac{1}{4}\right)^2 < \frac{abc}{4},$$

therefore

$$ab + bc + ca - \frac{1}{4} < \frac{\sqrt{abc}}{2},$$

hence

$$ab + bc + ca < \frac{\sqrt{abc}}{2} + \frac{1}{4}.$$

Proof (Second proof). Since $abc > 0$, there are one or three positive numbers among a, b, and c. In the first case, we may assume $a > 0$, $b, c < 0$, then

$$ab + bc + ca = b(a + c) + ca < b(a + c) = b(1 - b) < 0,$$

and the inequality in question clearly holds.

Now we assume $a, b, c > 0$, and $a \geq b \geq c$, so $a \geq \dfrac{1}{3}$, $0 < c \leq \dfrac{1}{3}$. We have

$$ab + bc + ca - \frac{\sqrt{abc}}{2} = c(a + b) + \sqrt{ab}\left(\sqrt{ab} - \frac{\sqrt{c}}{2}\right)$$

$$= c(1 - c) + \sqrt{ab}\left(\sqrt{ab} - \frac{\sqrt{c}}{2}\right).$$

Since $\sqrt{ab} \geq \sqrt{\dfrac{b}{3}} \geq \sqrt{\dfrac{c}{3}} > \dfrac{\sqrt{c}}{2}$ and $\sqrt{ab} \leq \dfrac{a + b}{2} = \dfrac{1 - c}{2}$,

$$c(1 - c) + \sqrt{ab}\left(\sqrt{ab} - \frac{\sqrt{c}}{2}\right) \leq c(1 - c) + \frac{1 - c}{2}\left(\frac{1 - c}{2} - \frac{\sqrt{c}}{2}\right)$$

$$= \frac{1}{4} - \frac{3c^2}{4} + \frac{c\sqrt{c}}{4} + \frac{c}{2} - \frac{\sqrt{c}}{4}.$$

So it is enough to show that

$$\frac{3c^2}{4} - \frac{c\sqrt{c}}{4} - \frac{c}{2} + \frac{\sqrt{c}}{4} > 0,$$

or, since $c > 0$, equivalently,

$$3c\sqrt{c} - c - 2\sqrt{c} + 1 > 0. \qquad ③$$

Note that $0 < c \leq 1/3$, so

$$\frac{1}{3} - c \geq 0. \qquad\qquad ④$$

By the AM-GM inequality,

$$3c\sqrt{c} + \frac{1}{3} + \frac{1}{3} \geq 3 \left(3c\sqrt{c} \cdot \frac{1}{3} \cdot \frac{1}{3} \right)^{1/3} = 9^{1/3}\sqrt{c} > 2\sqrt{c}. \qquad ⑤$$

adding ④ and ⑤ we get ③.

2 (40 marks) As in Figure 2.1, in an acute triangle ABC, $\angle BAC \neq 60°$. The lines BD and CE are tangent to the excircle of the triangle ABC, and $BD = CE = BC$. Line DE meets the extensions of AB and AC at F and G, respectively. Let M be the intersection point of CF and BD, and N be the intersection point of CE and BG. Prove that $AM = AN$.

Proof (First proof). As in Figure 2.2, let K be the intersection of the two tangent lines BD and CE, $BK = CK$. By $BD = CE$, we get $DE \parallel BC$. Draw the bisector of $\angle BAC$ and meet BC at L, connect L to M and N.

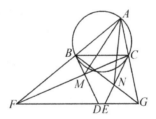

Fig. 2.1

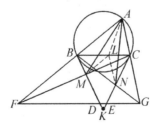

Fig. 2.2

Since $DE \parallel BC$, $\angle ABC = \angle DFB$, $\angle FDB = \angle DBC = \angle BAC$, so $\triangle ABC$ is similar to $\triangle DFB$. From this and the fact $DE \parallel BC$, $BD = BC$, and the angle bisector theorem,

$$\frac{MC}{MF} = \frac{BC}{FD} = \frac{BD}{FD} = \frac{AC}{AB} = \frac{LC}{LB}.$$

We get $LM \parallel BF$.

Similarly, $LN \parallel CG$. Hence,

$$\angle ALM = \angle ALB + \angle BLM = \angle ALB + \angle ABL$$

$$= 180° - \angle BAL = 180° - \angle CAL$$

$$= \angle ALC + \angle ACL = \angle ALC + \angle CLN = \angle ALN.$$

By $BC \parallel FG$ and the angle bisector theorem,

$$\frac{LM}{LN} = \frac{LM}{BF} \cdot \frac{BF}{CG} \cdot \frac{CG}{LN} = \frac{CL}{BC} \cdot \frac{AB}{AC} \cdot \frac{BC}{BL} = \frac{CL}{BL} \cdot \frac{AB}{AC} = 1.$$

Hence $LM = LN$.

Now $\triangle ALM \cong \triangle ALN$, since AL is the common edge, $\angle ALM = \angle ALN$, and $LM = LN$. $AM = AN$ follows. $\qquad\square$

Proof (Second proof). $\angle DBC = \angle BAC = \angle ECB$, because BD and EC are tangent to the circle; Together with the fact $BD = CE$, we deduce that $BCED$ is an isosceles trapezoid, and $DE \parallel BC$.

$\angle BFD = \angle ABC$, $\angle FDB = \angle DBC = \angle BAC$, so $\triangle DFB$ and $\triangle ABC$ are similar.

Denote the interior angles of $\triangle ABC$ as A, B, C, and the sides $BC = a$, $CA = b$, and $AB = c$.

$\triangle DFB$ and $\triangle ABC$, so we get $FD : c = BD : b = a : b$, then $FD = ac/b$.

$BC \parallel FD$, so $BM : MD = BC : FD = b : c$; $BD = a$, and we get

$$BM = \frac{ab}{b+c}. \qquad \qquad ①$$

In $\triangle ABM$, $\angle ABM = B + A$. By the law of cosines,

$$AM^2 = c^2 + \frac{a^2 b^2}{(b+c)^2} - \frac{2abc}{b+c} \cos(A+B)$$

$$= c^2 + \frac{a^2 b^2}{(b+c)^2} - \frac{2abc}{b+c} \frac{a^2 + b^2 - c^2}{2ab}$$

$$= \frac{1}{(b+c)^2}[c^2(b+c)^2 + a^2b^2 + c(a^2 + b^2 - c^2)(b+c)]$$

$$= \frac{1}{(b+c)^2}(b^2c^2 + 2bc^3 + c^4 + a^2b^2$$

$$+ a^2bc + a^2c^2 + b^3c + b^2c^2 - bc^3 - c^4)$$

$$= \frac{1}{(b+c)^2}(2b^2c^2 + bc^3 + b^3c + a^2b^2 + a^2c^2 + a^2bc) \qquad ②$$

Express CN and AN^2 in the same way as we did for BM and AM^2, we only need to interchange b and c in the calculation in ① and ②. The final expression in ② is symmetric about b and c, so $AM^2 = AN^2$, then $AM = AN$. □

③ (50 marks) Let $S = \{1, 2, 3, \ldots, 100\}$. Determine the maximum integer k such that there exist k distinct non-empty subsets of S satisfy: for any two distinct subsets among the k subsets, if they intersect, the minimum element in their intersection equals none of the maximum elements of the two subsets.

Solution Denote $\min A$ (resp. $\max A$) the minimum (resp. maximum) element of any finite non-empty set of real numbers A.

Consider all the subsets of S that contains the element 1 and that are of size at least 2. There are $2^{99} - 1$ such subsets, and they satisfy the requirement since for any two of them, the minimum element of their intersection is 1, and equals neither of the maximum element of the two subsets. So $k_{\max} \geq 2^{99} - 1$.

Now we prove that when $k \geq 2^{99}$, no k subsets can satisfy the requirement. We prove by induction: For integers $n \geq 3$, when $m \geq 2^{n-1}$, among any m distinct subsets $A_1, A_2, \ldots, A_m$ of $\{1, 2, \ldots, n\}$, there exist $i \neq j$ such that

$$A_i \cap A_j \neq \emptyset \quad \text{and} \quad \min(A_i \cap A_j) = \max A_i. \qquad ①$$

Clearly we only need to prove the proposition for $m = 2^{n-1}$.

When $n = 3$, divide the non-empty subsets of $\{1, 2, 3\}$ into 3 groups:

$$\{3\}, \{1, 3\}, \{2, 3\};$$

$$\{2\}, \{1, 2\};$$

$$\{1\}, \{1, 2, 3\}.$$

By the pigeonhole principle, among any 4 non-empty subsets, two of them, A_i and A_j are from the same group, where A_i occurs before A_j in our list, and it is easy to check that they satisfy ①.

Suppose our proposition holds for n ($n \geq 3$), and consider the situation for $n + 1$. Let $\mathcal{F} = \{A_1, A_2, \ldots, A_{2^n}\}$ be a collection of 2^n subsets of $\{1, 2, \ldots, n+1\}$. If there are 2^{n-1} subsets in $\mathcal{F}$ do not contain the element $n + 1$, by inductive hypothesis, two sets among the 2^{n-1} satisfy ①.

In the other case, at most $2^{n-1} - 1$ subsets in $\mathcal{F}$ do not contain $n + 1$, so $n + 1$ is contained in at least $2^{n-1} + 1$ subsets in $\mathcal{F}$. For each of these subsets, remove the element $n + 1$ and we get a collection $\mathcal{F}'$ of $2^{n-1} + 1$ subsets of $\{1, 2, \ldots, n\}$.

Divide all the subsets of $\{1, 2, \ldots, n\}$ into 2^{n-1} pairs, each subset is paired with its complement. By pigeonhole principle, two subsets in $\mathcal{F}'$ are from the same pair, i.e., they are complement to each other. The two corresponding subsets, A_i and A_j, in $\mathcal{F}$ has $A_i \cap A_j = \{n + 1\}$ and clearly satisfy ①. So our proposition holds for $n + 1$.

We conclude that, for our problem, $k_{\max} = 2^{99} - 1$. $\qquad \square$

4 (50 marks) Suppose integers $x_1, x_2, \ldots, x_{2014}$ are pairwise non-congruent modulo 2014; and integers $y_1, y_2, \ldots, y_{2014}$ are also pairwise non-congruent modulo 2014. Prove that, there is a permutation $z_1, z_2, \ldots, z_{2014}$ of $y_1, y_2, \ldots, y_{2014}$, so that the numbers $x_1 + z_1, x_2 + z_2, \ldots, x_{2014} + z_{2014}$ are pairwise non-congruent modulo 4028.

Proof. Let $k = 1007$. Without loss of generality, $x_i \equiv y_i \equiv i \pmod{2k}$, $1 \leq i \leq 2k$.

For each integer i, where $1 \leq i \leq k$, we set $z_i = y_i$ and $z_{i+k} = y_{i+k}$ if $x_i + y_i \not\equiv x_{i+k} + y_{i+k} \pmod{4k}$; otherwise set $z_i = y_{i+k}$ and $z_{i+k} = y_i$.

In the former case,

$$x_i + z_i = x_i + y_i \not\equiv x_{i+k} + y_{i+k} = x_{i+k} + z_{i+k} \pmod{4k}.$$

In the latter case, we also claim

$$x_i + z_i = x_i + y_{i+k} \not\equiv x_{i+k} + y_i = x_{i+k} + z_{i+k} \pmod{4k}.$$

Otherwise, we would have

$$x_i + y_i \equiv x_{i+k} + y_{i+k} \pmod{4k},$$

$$x_i + y_{i+k} \equiv x_{i+k} + y_i \pmod{4k}.$$

Add the above two together, we get $2x_i \equiv 2x_{i+k} \pmod{4k}$, then $x_i \equiv x_{i+k} \pmod{2k}$, contradicts the fact that $x_1, x_2, \ldots, x_{2014}$ are pairwise non-congruent modulo $2014(= 2k)$.

It is clear that $z_1, z_2, \ldots, z_{2k}$ constructed above is a permutation of $y_1, y_2, \ldots, y_{2k}$. Denote $w_i = x_i + z_i$ for $i = 1, 2, \ldots, 2k$. We now verify that $w_1, w_2, \ldots, w_{2k}$ are non-congruent modulo $4k$. It is enough to show that, (*) for any integer i, j where $1 \leq i < j \leq k$, the remainders of $w_i, w_j, w_{i+k}, w_{j+k}$ modulo $4k$ are all different.

Note that, in the construction, we already proved that

$$w_i \not\equiv w_{i+k} \pmod{4k}, \quad w_i \not\equiv w_{i+k} \pmod{4k}. \qquad \text{(1)}$$

We discuss 3 cases.

Case 1. $z_i = y_i$ and $z_j = y_j$. From our construction,

$$w_i \equiv w_{i+k} \equiv 2i \pmod{2k}, \quad w_j \equiv w_{j+k} \equiv 2j \pmod{2k}.$$

Since $2i \not\equiv 2j \pmod{2k}$, it is easy to see that w_i has different remainder modulo (even to) $2k$ than w_j and w_{j+k}, w_{i+k} has different remainder modulo $2k$ than w_j and w_{j+k}. Combined with (1), (*) is proved.

Case 2. $z_i = y_{i+k}$ and $z_{i+k} = y_i$. From our construction,

$$w_i \equiv w_{i+k} \equiv 2i + k \pmod{2k}, \quad w_j \equiv w_{j+k} \equiv 2j + k \pmod{2k}.$$

Similar to Case 1, w_i has different remainder modulo $2k$ than w_j and w_{j+k}, w_{i+k} has different remainder modulo $2k$ than w_j and w_{j+k}. Combined with (1), (*) is proved.

Case 3. $z_i = y_i$ and $z_j = y_{j+k}$. (The other case where $z_i = y_{i+k}$ and $z_j = y_i$ is symmetric to this.) By our construction

$$w_i \equiv w_{i+k} \equiv 2i \pmod{2k}, \quad w_j \equiv w_{j+k} \equiv 2j + k \pmod{2k}.$$

Since k is odd, $2i \not\equiv 2j + k \pmod{2}$ therefore $2i \not\equiv 2j + k \pmod{2k}$, so we still conclude that w_i has different remainder modulo $2k$ than w_j and w_{j+k}, w_{i+k} has different remainder modulo $2k$ than w_j and w_{j+k}. Thus (*) is proved. $\qquad \square$

China Mathematical Competition (Complementary Test)

1 (40 marks) Given real numbers $a_1, a_2, \ldots, a_n$, where $n \geq 2$. Prove that we may pick $\epsilon_1, \epsilon_2, \ldots, \epsilon_n \in \{1, -1\}$ so that

$$\left(\sum_{i=1}^{n} a_i\right)^2 + \left(\sum_{i=1}^{n} \epsilon_i a_i\right)^2 \leq (n+1)\left(\sum_{i=1}^{n} a_i^2\right)$$

Proof (First proof). We prove that

$$\left(\sum_{i=1}^{n} a_i\right)^2 + \left(\sum_{i=1}^{\lfloor n/2 \rfloor} a_i - \sum_{j=\lfloor n/2 \rfloor+1}^{n} a_j\right)^2 \leq (n+1)\left(\sum_{i=1}^{n} a_i^2\right). \qquad \textcircled{1}$$

In other words, in order to satisfy the inequality in question, we may just take $\epsilon_i = 1$ for $i = 1, 2, \ldots, \lfloor n/2 \rfloor$ and $\epsilon_i = -1$ for $i = \lfloor n/2 \rfloor + 1, \ldots, n$.

We start from the left hand side of $\textcircled{1}$, it is

$$\left(\sum_{i=1}^{\lfloor n/2 \rfloor} a_i + \sum_{j=\lfloor n/2 \rfloor+1}^{n} a_j\right)^2 + \left(\sum_{i=1}^{\lfloor n/2 \rfloor} a_i - \sum_{j=\lfloor n/2 \rfloor+1}^{n} a_j\right)^2$$

$$= 2\left(\sum_{i=1}^{\lfloor n/2 \rfloor} a_i\right)^2 + 2\left(\sum_{j=\lfloor n/2 \rfloor+1}^{n} a_j\right)^2$$

$$\leq 2 \lfloor n/2 \rfloor \left(\sum_{i=1}^{\lfloor n/2 \rfloor} a_i^2 \right) + 2 \left(n - \lfloor n/2 \rfloor \right) \left(\sum_{j=\lfloor n/2 \rfloor + 1}^{n} a_j^2 \right)$$

(Cauchy-Schwarz)

$$= 2 \lfloor n/2 \rfloor \left(\sum_{i=1}^{\lfloor n/2 \rfloor} a_i^2 \right) + 2 \left(\left\lfloor \frac{n+1}{2} \right\rfloor \right) \left(\sum_{j=\lfloor n/2 \rfloor + 1}^{n} a_j^2 \right)$$

$$\leq n \left(\sum_{i=1}^{\lfloor n/2 \rfloor} a_i^2 \right) + (n+1) \left(\sum_{j=\lfloor n/2 \rfloor + 1}^{n} a_j^2 \right)$$

$$\leq (n+1) \left(\sum_{i=1}^{n} a_i^2 \right).$$

Thus, ① is proved.   □

Proof (Second proof). First of all, by symmetry, we may assume $a_1 \geq a_2 \geq \cdots \geq a_n$. Furthermore, if we change all the negative a_i's to their absolute values, the $\left(\sum_{i=1}^{n} a_i \right)^2$ term on the left hand side of the inequality in question does not decrease, the $\left(\sum_{i=1}^{n} a_i^2 \right)$ on the right hand side is unchanged. Once we have a good choice of the ϵ_i's for the modified a_i's, we may change the signs of the ϵ_i's that correspond to the modified a_i's and get a good assignment for the original problem. So we may further assume $a_1 \geq a_2 \geq \cdots \geq a_n \geq 0$.

Lemma 4.1 *Suppose $a_1 \geq a_2 \geq \cdots \geq a_n \geq 0$, then $0 \leq \sum_{i=1}^{n}(-1)^{i-1}a_i \leq a_1$.*

Proof (of lemma). Note that $a_i \geq a_{i+1}$ for each $i = 1, 2, \ldots, n-1$.
  When n is even,

$$\sum_{i=1}^{n}(-1)^{i-1}a_i = (a_1 - a_2) + (a_3 - a_4) + \cdots + (a_{n-1} - a_n) \geq 0,$$

$$\sum_{i=1}^{n}(-1)^{i-1}a_i = a_1 - (a_2 - a_3) - \cdots - (a_{n-2} - a_{n-1}) - a_n \leq a_1;$$

and when n is odd,

$$\sum_{i=1}^{n}(-1)^{i-1}a_i = (a_1 - a_2) + (a_3 - a_4) + \cdots + (a_{n-1} - a_n) + a_n \geq 0,$$

$$\sum_{i=1}^{n}(-1)^{i-1}a_i = a_1 - (a_2 - a_3) - \cdots - (a_{n-1} - a_n) \leq a_1.$$

This completes the proof of the lemma.

Now, by Cauchy-Schwarz and the above lemma,

$$\left(\sum_{i=1}^{n}a_i\right)^2 + \left(\sum_{i=1}^{n}(-1)^{i-1}a_i\right)^2 \leq n\left(\sum_{i=1}^{n}a_i^2\right) + a_1^2 \leq (n+1)\left(\sum_{i=1}^{n}a_i^2\right).$$

In other words, we may pick $\epsilon_i = (-1)^{i-1}$ and the inequality in question is satisfied. $\qquad\square$

2. (40 marks) Let $\mathcal{S} = \{A_1, A_2, \ldots, A_n\}$, where $A_i, A_2, \ldots, A_n$ are n distinct finite sets ($n \geq 2$), and such that $A_i \cup A_j \in \mathcal{S}$ whenever $A_i, A_j \in \mathcal{S}$. Let $k = \min_{1 \leq i \leq n}|A_i|$ and suppose that $k \geq 2$. Prove that there exists an element $x \in \cup_{i=1}^{n}A_i$ which belongs to at least n/k of the A_i's.

Proof. We may assume $|A_1| = k$. Suppose s of the sets in $\mathcal{S}$ are disjoint from A_1, denote them by $B_1, B_2, \ldots, B_s$. Suppose t sets in $\mathcal{S}$ are super-sets of A_1, denote them by $C_1, C_2, \ldots, C_t$. From the condition on $\mathcal{S}$, $B_i \cup A_1 \in \mathcal{S}$, so $B_i \cup A_1 \in \{C_1, \ldots, C_t\}$. Thus we have a mapping

$$f : \{B_1, B_2, \ldots, B_s\} \to \{C_1, C_2, \ldots, C_t\}, f(B_i) = B_i \cup A_1.$$

It is easy to see that f is injective, so $s \leq t$.

Write the elements of A_1 as $a_1, a_2, \ldots, a_k$. Consider $\mathcal{F}$, the collection of the $n - s - t$ sets in $\mathcal{S}$ other than the B_i's and the C_i's. Denote by x_i the number of sets in $\mathcal{F}$ that contains a_i. Since each set in $\mathcal{F}$ has a non-empty intersection with A_1, so it contains at least one of the a_i's. It follows that

$$x_1 + x_2 + \cdots + x_n \geq n - s - t.$$

Without loss of generality, $x_1 = \max_{1 \leq i \leq k} x_i$, so $x_1 \geq \dfrac{n-s-t}{k}$, i.e., a_1 belongs to at least $\dfrac{n-s-t}{k}$ sets in $\mathcal{F}$. Noting that $A_1 \subseteq C_i$ for each $i = 1, 2, \ldots, t$, so each of the C_i's contains a_1. So, the number of sets in $\mathcal{S}$

containing a_1 is at least

$$\frac{n-s-t}{k}+t=\frac{n-s+(k-1)t}{k}\geq\frac{n-s+t}{k}\geq\frac{n}{k}.$$

Here we used the facts that $k\geq 2$ and $t\geq s$.   □

3 (50 marks) As in the figure, $\triangle ABC$ is inscribed in the circle centered at O, P is a point on $\overset{\frown}{BC}$, K is a point on the segment AP so that BK bisects $\angle ABC$. The circle Γ passing through K, P, and C intersects AC at D. Draw BD and it intersects Γ at a point E; draw PE and extend it to meet AB at a point F.
Prove that $\angle ABC = 2\angle FCB$.

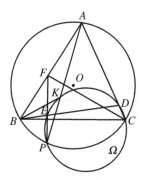

Fig. 3.1

Proof (First proof). As in Figure 3.2, let L be the other intersection point of CF and the circle Γ, draw PB, PC, BL, and KL.

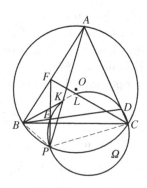

Fig. 3.2

Note that now C, D, L, K, E, and P are all on the circle Γ. Together with the fact that A, B, P, and C are concyclic, we have

$$\angle FEB = \angle DEP = 180° - \angle DCP = \angle ABP = \angle FBP,$$

so $\triangle FBE \sim \triangle FPB$, therefore $FB^2 = FE \cdot FP$.

By the power of a point theorem, $FE \cdot FP = FL \cdot FC$, so

$$FB^2 = FL \cdot FC,$$

therefore $\triangle FBL \sim \triangle FCB$.

Hence

$$\angle FLB = \angle FBC = \angle APC = \angle KPC = \angle FLK,$$

i.e., B, K, and L are collinear.

Now, since $\triangle FBL \sim \triangle FCB$,

$$\angle FCB = \angle FBL = \angle FBE = \frac{1}{2}\angle ABC,$$

i.e., $\angle ABC = 2\angle FCB$. $\qquad\qquad\square$

Proof (Second proof). As in Figure 3.3, let L be the other intersection point of CF and the circle Γ, draw PC and KL. Apply Pascal's theorem to the generalized inscribed hexagon $DCLKPE$, A — the intersection of DC and KP, F — the intersection of CL and PE, and B' — the intersection of LK and ED, are collinear. So B' is the intersection of AF and ED, which is B. Hence, B, K, and L are collinear.

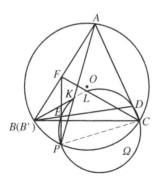

Fig. 3.3

Note that A, B, P, and C are concyclic, and L, K, P, and C are concyclic, so

$$\angle ABC = \angle APC = \angle FLK = \angle FCB + \angle LBC.$$

Because BK bisects $\angle ABC$, we have $\angle LBC = \dfrac{1}{2}\angle ABC$, therefore $\angle ABC = 2\angle FCB$.   □

④ (50 marks) Determine all the positive integers k satisfying: $2^{(k-1)n+1}$ does not divide $\dfrac{(kn)!}{n!}$ for any positive integer n.

Solution For positive integer m, $\nu_2(m)$ is the standard 2-adic valuation of m, i.e., the exponent of 2 in the standard factorization of m; and denote by $S(m)$ the sum of the digits in the binary representation of m. It is well known that

$$\nu_2(m!) = m - S(m). \qquad \text{①}$$

The condition that $2^{(k-1)n+1}$ does not divide $\dfrac{(kn)!}{n!}$ is equivalent to

$$\nu_2\left(\frac{(kn)!}{n!}\right) \le (k-1)n,$$

which in turn is equivalent to

$$kn - \nu_2((kn)!) \ge n - \nu_2(n!).$$

By ①, we are seeking all the positive integers k satisfying: $S(kn) \ge S(n)$ for all positive integers n.

We prove that they are precisely the powers of 2, i.e., 2^a for $a = 0, 1, \ldots$.

On one hand, $S(2^a n) = S(n)$ for all positive integers n, so $k = 2^a$ satisfied our requirement.

On the other hand, suppose k is not a power of 2, let $k = 2^a \cdot q$, where $a \ge 0$ and q is an odd number bigger than 1. We construct a positive integer n such that $S(kn) < S(n)$. Since $S(kn) = S(2^a qn) = S(qn)$, so equivalently we only need to construct an m such that q divides m and

$$S(m) < S(m/q).$$

Because $(2, q) = 1$, there exists positive integer u such that $2^u \equiv 1 \pmod{q}$ — in fact we may take $u = \phi(q)$ by Euler's theorem.

Suppose the binary representation of q is $2^{a_1} + 2^{a_2} + \cdots + 2^{a_t}$, $0 = a_1 < a_2 < \cdots < a_t$, $t \ge 2$.

Take $m = 2^{a_1} + 2^{a_2} + \cdots + 2^{a_{t-1}} + 2^{a_t + tu}$, then $S(m) = t$, and

$$m = q + 2^{a_t}(2^{tu} - 1) \equiv 0 \pmod{q}.$$

We have

$$\frac{m}{q} = 1 + 2^{a_t} \cdot \frac{2^{tu} - 1}{q}$$

$$= 1 + 2^{a_t} \cdot \frac{2^u - 1}{q}(1 + 2^u + \cdots + 2^{(t-1)u})$$

$$= 1 + \sum_{l=0}^{t-1} \frac{2^u - 1}{q} \cdot 2^{lu + a_t}. \qquad \textcircled{2}$$

Since $0 < \dfrac{2^u - 1}{q} < 2^u$, so the highest term in the binary representation of $\dfrac{2^u - 1}{q}$ is less than u. It follows that, for any i and j where $0 \le i < j \le t-1$, the binary representations of $\dfrac{2^u - 1}{q} \cdot 2^{iu + a_t}$ and $\dfrac{2^u - 1}{q} \cdot 2^{ju + a_t}$ do not share any common terms.

Also, $a_t > 0$, so, for any $1 \le l \le t - 1$, the binary representation of $\dfrac{2^u - 1}{q} \cdot 2^{lu + a_t}$ does not contain the term 1. Hence, by $\textcircled{2}$,

$$S\left(\frac{m}{q}\right) = 1 + S\left(\frac{2^u - 1}{q}\right) \cdot t > t = S(m).$$

In conclusion, the numbers we seek are $k = 2^a$, $a = 0, 1, 2, \ldots$. $\qquad \square$

China Mathematical Olympiad

$$2014 \text{ (Chongqing)}^*$$

First Day

December 20, 2014

8:00–12:30

1. Given a real number $r \in (0, 1)$. Prove that if n complex numbers $z_1, z_2, \ldots, z_n$ satisfy $|z_k - 1| \le r$ for $k = 1, 2, \ldots, n$, then

$$|z_1 + z_2 + \cdots + z_n| \left| \frac{1}{z_1} + \frac{1}{z_2} + \cdots + \frac{1}{z_n} \right| \ge n^2 (1 - r^2).$$

Proof. Let $z_k = x_k + y_k i$, $x_k, y_k \in \mathbb{R}$, $k = 1, 2, \ldots, n$. First we show that

$$\frac{x_k^2}{x_k^2 + y_k^2} \ge 1 - r^2, \quad k = 1, 2, \ldots, n. \qquad \text{(1)}$$

Write $u = \dfrac{x_k^2}{x_k^2 + y_k^2}$. Since $|x_k - 1| \le r < 1$, we have $x_k > 0$, hence $u > 0$. Also $y_k^2 = \left(\dfrac{1}{u} - 1 \right) x_k^2$, it follows that

$$r^2 \ge |z_k - 1|^2 = (x_k - 1)^2 + \left(\frac{1}{u} - 1 \right) x_k^2 = \frac{1}{u}(x_k - u)^2 + 1 - u \ge 1 - u,$$

hence $u \ge 1 - r^2$, i.e., inequality (1) holds.

*The China Mathematical Olympiad, organized by the China Mathematical Olympiad Committee, is held in January every year. About 150 winners of the China Mathematical Competition lasts for two days, and there are three problems to be completed within 4.5 hours each day.

Since

$$|z_1 + z_2 + \cdots + z_n| \geq |\mathrm{Re}(z_1 + z_2 + \cdots + z_n)| = \sum_{k=1}^{n} x_k,$$

and $\dfrac{1}{z_k} = \dfrac{x_k - y_k i}{x_k^2 + y_k^2}$, $k = 1, 2, \ldots, n$, we have

$$\left| \frac{1}{z_1} + \frac{1}{z_2} + \cdots + \frac{1}{z_k} \right| \geq \left| \mathrm{Re}\left(\frac{1}{z_1} + \frac{1}{z_2} + \cdots + \frac{1}{z_k} \right) \right| = \sum_{k=1}^{n} \frac{x_k}{x_k^2 + y_k^2}.$$

Note that $x_k > 0$ for $k = 1, 2, \ldots, n$. By Cauchy's inequality, we have

$$|z_1 + z_2 + \cdots + z_n| \cdot \left| \frac{1}{z_1} + \frac{1}{z_2} + \cdots + \frac{1}{z_n} \right|$$

$$\geq \left(\sum_{k=1}^{n} x_k \right) \left(\sum_{k=1}^{n} \frac{x_k}{x_k^2 + y_k^2} \right)$$

$$\geq \left(\sum_{k=1}^{n} \sqrt{\frac{x_k^2}{x_k^2 + y_k^2}} \right)^2 \geq (n\sqrt{1 - r^2})^2 = n^2(1 - r^2). \qquad \square$$

2 As shown in Figure 2.1, points A, B, D, E, F, C lie on a circle with $AB = AC$. Lines AD and BE intersect at point P; lines AF and CE intersect at point R; lines BF and CD intersect at point Q; lines AD and BF intersect at point S; lines AF and CD intersect at point T. Let K be a point on the segment ST such that $\angle SKQ = \angle ACE$. Prove that $\dfrac{SK}{KT} = \dfrac{PQ}{QR}$.

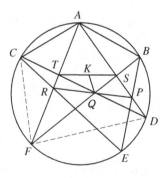

Fig. 2.1

Proof. Since $AB = AC$, it follows that $\angle ADC = \angle AFB$, hence S, D, F, T are concyclic, and then $\angle QSK = \angle TDF = \angle RAC$. Combining with the given condition $\angle SKQ = \angle ACE$, we see that $\triangle QSK \sim \triangle RAC$, and similarly $\triangle QTK \sim \triangle PAB$.

It follows that

$$\frac{SK}{KQ} = \frac{AC}{CR}, \quad \frac{KQ}{KT} = \frac{BP}{BA},$$

hence

$$\frac{SK}{KT} = \frac{SK}{KQ} \cdot \frac{KQ}{KT} = \frac{AC}{CR} \cdot \frac{BP}{BA} = \frac{BP}{CR}. \qquad ①$$

By Pascal's theorem, $P, Q,$ and R are collinear. Let J be a point on the ray CD such that $\triangle BCJ \sim \triangle BAP$. Since $\dfrac{BP}{BJ} = \dfrac{AB}{CB}$, and

$$\angle ABC = \angle PBA - \angle PBC = \angle JBC - \angle PBC = \angle JBP,$$

we get $\triangle BPJ \sim \triangle BAC$, which implies $PB = PJ$ once noticing $AB = AC$. Also note that $\angle DPE = \angle BPA = \angle BJC$, it follows that B, J, D, P are concyclic, therefore

$$\angle PJQ = \angle DBE = \angle DCE,$$

thus $PJ \parallel ICR$. As a consequence, we have

$$\frac{BP}{CR} = \frac{PJ}{CR} = \frac{PQ}{QR}. \qquad ②$$

By equations ① and ②, we proved the result. $\qquad \square$

3 Given integer $n \geq 5$. Find the smallest integer m such that there exist two sets of integers A, B satisfying both following conditions:

(1) $|A| = n$, $|B| = m$, and $A \subseteq B$;

(2) For any two distinct elements x, y in B, we have $x + y \in B$ if and only if $x, y \in A$.

Solution The answer is $m = 3n - 3$.

First, let N be an integer $\geq 2n$. Put

$$A = \{N + 1, N + 2, \ldots, N + n\},$$

$$B = A \cup \{2N + 3, 2N + 4, \ldots, 2N + 2n - 1\}.$$

Clearly A and B satisfy condition (1). For any $x, y \in A$, $x \neq y$, since $2N + 3 \leq x + y \leq 2N + 2n - 1$, we have $x + y \in B$. On the other hand, for any $x, y \in B$, $x \neq y$, and x, y do not both belong to A, we have $x + y \geq 3N > 2N + 2n - 1$, hence $x + y \notin B$. Thus, condition (2) also holds for A, B. We see that there exist sets A, B with required properties for $m = 3n - 3$.

Next we show that for any two sets A, B satisfying conditions (1) and (2), $|B| \geq 3n - 3$. We may assume without loss of generality that the number of positive elements in A is not less than the number of negative elements in A, otherwise we may replace A and B with $-A$ and $-B$ respectively, then $-A$ and $-B$ also satisfy conditions (1) and (2), and the number of positive elements in $-A$ is not less than the number of negative elements in $-A$.

Write the elements of A as $a_1 < a_2 \cdots < a_n$. Since $n \geq 5$, there are at least 2 positive numbers in A, i.e. $a_{n-1} > 0$. It follows that $a_{n-1} + a_n > a_n$, thus $a_{n-1} + a_n \in B \backslash A$, and $|B| \geq n + 1$. We claim that $0 \notin A$. If on the contrary $0 \in A$, take arbitrary $x \in B \backslash A$, then $0 + x = x \in B$. But $0 \in A$ and $x \notin A$, so it violates condition (2). Since $n \geq 5$ and $0 \notin A$, so A has at least 3 positive numbers, i.e. $a_{n-2} > 0$. It follows that $a_n + a_{n-2} > a_n$, thus $a_n + a_{n-2} \notin A$.

We now prove two conclusions.

(1) For any integer i, $1 \leq i < n$, we have $a_n + a_i \in B \backslash A$.

Assume on the contrary that $a_n + a_i \in A$ for some i, say $a_n + a_i = a_j$, then $a_i < a_j < a_n$. If $j = n - 1$, then

$$(a_i + a_n) + a_{n-2} = a_{n-1} + a_{n-2} \in B.$$

But $a_i \in A$ and $a_n + a_{n-2} \in B \backslash A$, this violates condition (2). If $j < n - 1$, then

$$(a_i + a_n) + a_{n-1} = a_j + a_{n-1} \in B.$$

But $a_i \in A$ and $a_n + a_{n-1} \in B \backslash A$, this again violates condition (2). This completes the proof of conclusion (1).

(2) For any integer i, $2 \leq i < n$, we have $a_1 + a_i \in B \backslash A$.

Assume on the contrary that $a_1 + a_i \in A$ for some i, say $a_1 + a_i = a_j$. If $j \neq n$, then

$$(a_1 + a_i) + a_n = a_j + a_n \in B.$$

But $a_1 \in A$, and $a_i + a_n \in B \backslash A$ by conclusion (1), this violates condition (2). So we must have $a_1 + a_i = a_n$. Since $n \geq 5$, let k be an integer such that $1 < k < n$ and $k \neq i$, it follows that $a_i + a_k > a_1 + a_k = a_n$, hence $a_i + a_k \in B \backslash A$. Note that

$$(a_1 + a_i) + a_k = a_n + a_k \in B.$$

But $a_1 \in A$ and $a_i + a_k \in B \backslash A$, this again violates condition (2). This completes the proof of conclusion (2).

By conclusions (1) and (2), the following $2n - 3$ pairwise distinct numbers

$$a_1 + a_2 < a_1 + a_3 < \cdots < a_1 + a_n < a_2 + a_n < \cdots < a_{n-1} + a_n$$

all belong to $B \backslash A$, thus $|B| \geq 3n - 3$. This proves our result. $\qquad \square$

Second Day

December 21, 2014

8:00–12:30

4   Find all integer k with the following property: there exist infinitely many positive integers n such that $n + k$ does not divide C_{2n}^n.

Solution   We claim that all integers except 1 have the required property. First we show that $k = 1$ does not have the required property. In fact

$$\frac{1}{n+1}C_{2n}^n = \frac{(2n)!}{n!(n+1)!}((n+1) - n) = C_{2n}^n - C_{2n}^{n-1}$$

is the difference of two binomial coefficients, thus an integer. So $n + 1 \mid C_{2n}^n$ for all positive integers n.

For $k > 1$, let p be a prime divisor of k. Take $n = p^m - k$ for sufficiently large m such that $n > 0$. Clearly there are infinitely many such n. We then show that $n + k \nmid C_{2n}^n$ for such n, i.e., $p^m \nmid C_{2n}^n$. Denote the exponent of p dividing $C_{2n}^n = \dfrac{(2n)!}{(n!)^2}$ to be α. We have,

$$\alpha = \sum_{l=1}^{\infty} \left[\frac{2n}{p^l}\right] - 2\sum_{l=1}^{\infty}\left[\frac{n}{p^l}\right]$$

$$= \sum_{l=1}^{m}\left(\left[\frac{2n}{p^l}\right] - 2\left[\frac{2n}{p^l}\right]\right) \text{ (since } 2n < 2p^m \le p^{m+1},$$

$$\text{so } \left[\frac{2n}{p^l}\right] = 0 \text{ for } l \ge m+1)$$

$$= \sum_{l=1}^{m}\left(\left(2p^{m-l} + \left[\frac{-2k}{p^l}\right]\right) - 2\left(p^{m-l} + \left[\frac{-k}{p^l}\right]\right)\right)$$

$$= \left[\frac{-2k}{p}\right] - 2\left[\frac{-k}{p}\right] + \sum_{l=2}^{m}\left(\left[\frac{-2k}{p^l}\right] - 2\left[\frac{-k}{p^l}\right]\right)$$

$$\le \left[\frac{-2k}{p}\right] - 2\left[\frac{-k}{p}\right] + \sum_{l=2}^{m}1 \text{ (by } [2x] - 2[x] \le 1)$$

$$= m - 1 < m \ \left(\text{since } p|k, \text{ so } \left[\frac{-2k}{p}\right] - 2\left[\frac{-k}{p}\right] = 0\right),$$

hence $p^m \nmid C_{2n}^n$.

For $k \leq 0$, take arbitrary odd prime number $p > 2|k|$, and let $n = p + |k|$. There are an amount of infinitely such n since there are infinite primes. We show that $n + k \nmid C_{2n}^n$ for such n, i.e. $p \nmid C_{2n}^n$. In fact,

$$C_{2n}^n = \frac{2n \times (2n - 1) \times \cdots \times (n + 1)}{n \times (n - 1) \times \cdots \times 1} \qquad \textcircled{1}$$

Note that $k \leq 0$ and $p > 2|k|$, we have $0 < n + k < n + 1 \leq 2(n + k) \leq 2n$ and $3(n + k) > 2n$, thus the exponents of p dividing the denominator and numerator of $\textcircled{1}$ are both 1 (appearing in $n + k$ and $2(n + k)$ respectively), hence $p \nmid C_{2n}^n$.

As a conclusion, all integers except 1 have the required property. $\qquad \square$

⑤ A group of 30 people participate in a meeting. It is known that each person has at most 5 friends in this group. Among any 5 people in this group, there exist two who are not friends. Find the maximum number k, such that in any situation there always exist k people in this group, among which no two are friends.

Solution The answer is $k = 6$.

We use 30 vertices to represent the 30 people, two vertices are adjacent if and only if their corresponding person are friends. Thus we obtain a simple graph G with 30 vertices satisfying the following conditions:

(i) the degree of each vertex of G is at most 5;
(ii) among any 5 vertices of G, there are two vertices which are nonadjacent.

Let V be the vertex set of G. We call a subset $A \subseteq V$ an *independent set* if the vertices in A are pairwise nonadjacent. An independent set with maximum cardinality is called a *maximum independent set*.

(1) We first show that a maximum independent set in G has cardinality at least 6.

Let X be a maximum independent set in G. Each vertex in $V \backslash X$ is adjacent to some vertex in X, otherwise if there is some vertex $a \in V \backslash X$ which is not adjacent to any vertex in X, then $X \cup \{a\}$ is an independent set, contradicting with the maximality of X. It follows that there are at least $|V \backslash X| = 30 - |X|$ edges between $V \backslash X$ and X. Also note that each vertex in X has at most 5 edges, thus

$$30 - |X| \leq 5|X|, \qquad \textcircled{1}$$

hence $|X| \geq 5$. If $|X| = 5$, that is inequality in ① is an equality, then the $30 - |X|$ edges distributed over the 5 vertices in X, and each vertex in X has exactly 5 neighbours in $V \backslash X$. Since $|V \backslash X| = 25$, it follows that the neighbourhood of vertices in X are disjoint. Let $X = \{a, b, c, d, e\}$, and the neighbourhood of a be $Y_a = \{y_1, y_2, y_3, y_4, y_5\}$. It follows from condition (ii) that there are two nonadjacent vertices in Y_a, say y_1 and y_2. However y_1, y_2 are not in the neighbourhood of any vertex of b, c, d, e, thus $\{y_1, y_2, b, c, d, e\}$ is an independent set of cardinality 6, a contradiction. This proves $|X| \geq 6$.

(2) Next we construct a graph G satisfying conditions (i) and (ii) whose maximum independent set has cardinality 6.

Divide V into 3 sets V_1, V_2, V_3 with $|V_i| = 10$, $i = 1, 2, 3$. Let

$$V_1 = \{A_1, A_2, A_3, A_4, A_5, B_1, B_2, B_3, B_4, B_5\},$$

the edges are connected in following way (as shown in the figure):

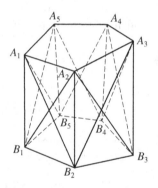

(i) join $A_i A_{i+1}$ for $i = 1, 2, 3, 4, 5$;
(ii) join $B_i B_{i+1}$ for $i = 1, 2, 3, 4, 5$;
(iii) join $A_i B_i$, $A_i B_{i+1}$ and $A_i B_{i-1}$ for $i = 1, 2, 3, 4, 5$,

where $A_6 = A_1$, $B_6 = B_1$, $B_0 = B_5$. Exactly the same graphs are constructed on sets V_2 and V_3, and for $1 \leq i < j \leq 3$, vertices between V_i and V_j are nonadjacent.

In this graph G, each vertex has degree 5, and there are two nonadjacent vertices among any 5 vertices.

Let X be a maximum independent set of G, we claim that $|V_1 \cap X| \leq 2$. Indeed, since A_i, A_{i+1} are adjacent ($i = 1, 2, 3, 4, 5$), at most two vertices of $A_1, \ldots A_5$ belong to X. Similarly, at most two vertices of $B_1, \ldots, B_5$ belong to X. If there are exactly two vertices of $A_1, \ldots, A_5$ belonging to X, we may assume without loss of generality that $A_1, A_3 \in A$. Note that each of $B_1, \ldots, B_5$ is adjacent to either A_1 or A_3, thus none of $B_1, \ldots, B_5$ lies in X. Similarly, if there are exactly two vertices of $B_1, \ldots, B_5$ belonging to X, then none of $A_1, \ldots, A_5$ lies in X, which shows that $|V_1 \cap X| \le 2$.

Similarly we obtain that $|V_2 \cap X| \le 2$, $|V_3 \cap X| \le 2$. Thus

$$|X| = |V \cap X| = |V_1 \cap X| + |V_2 \cap X| + |V_3 \cap X| \le 6.$$

It follows from part (1) that $|X| = 6$.

Combining (1) and (2), we see that the result is $k = 6$. $\qquad\square$

6 Let $a_1, a_2, \ldots$ be an infinite sequence of non-negative integers. Assume that for any positive integers m, n,

$$\sum_{i=1}^{2m} a_{in} \le m.$$

Prove that there exist positive integers k, d, such that

$$\sum_{i=1}^{2k} a_{id} = k - 2014.$$

Proof. Since $a_1, a_2, \ldots$ are nonnegative integers, we obtain by setting $m = 1$ in the assumption that $0 \le a_n + a_{2n} \le 1$ for any positive integer n. In particular, we have $a_n \in \{0, 1\}$ for every n, and $a_n = 0$ for infinitely many n.

Next we show that $a_n + a_{2n} = 0$ for infinitely many n. Assume on the contrary that $a_n + a_{2n} = 1$ for all except finitely many n, then there is a positive integer N, such that $a_n + a_{2n} = 1$ for all $n \ge N$. Since $a_n = 0$ for infinitely many n, fix a number $d > N$ such that $a_d = 0$. On one hand, by assumption, we have

$$\sum_{i=1}^{2m} a_{id} + \sum_{i=1}^{2m} a_{2id} \le m + m = 2m. \qquad \text{(1)}$$

On the other hand, since $id > N$ for all $i > 0$, we have

$$\sum_{i=1}^{2m} a_{id} + \sum_{i=1}^{2m} a_{2id} = \sum_{i=1}^{2m} (a_{id} + a_{2id}) = \sum_{i=1}^{2m} 1 = 2m. \qquad \text{(2)}$$

By (1) and (2), we obtain

$$\sum_{i=1}^{2m} a_{id} = m \qquad \text{(3)}$$

for all positive integers m. In particular,

$$a_{rd} + a_{2rd} = 1 \qquad \text{(4)}$$

for all $r \geq 1$. Put $r = 1, 2, 4$ in equation (4) respectively, we get $a_{2d} = 1 - a_d = 1$, $a_{4d} = 1 - a_{2d} = 0$ and $a_{8d} = 1 - a_{4d} = 1$. By (3) $(m = 2)$, $a_{3d} = 1$. By (4) $(m = 3, 6)$, we get $a_{6d} = 1 - a_{3d} = 0$, $a_{12d} = 1 - a_{6d} = 1$. Put $m - 3$ in (3), we get $a_{5d} = 1$, hence $a_{10d} = 1 - a_{5d} = 0$. Finally setting $m = 4, 5$ in (3) gives $a_{7d} = 0$ and $a_{9d} = 1$. It follows that

$$a_{3d} + a_{6d} + a_{9d} + a_{12d} = 1 + 0 + 1 + 1 = 3 < 2,$$

a contradiction with given assumption for $n = 3d$ and $m = 2$. Thus we proved that $a_n + a_{2n} = 0$ for infinitely many n.

Now let t be a positive integer such that there are at least 4028 positive integers $n \leq t$ satisfying $a_n + a_{2n} = 0$, then

$$\sum_{i=1}^{t}(a_i + a_{2i}) \leq t - 4028.$$

As a result, $\sum_{i=1}^{t} a_i \leq \dfrac{t}{2} - 2014$ or $\sum_{i=1}^{t} a_{2i} \leq \dfrac{t}{2} - 2014$. In any case, there exist positive integers t and d (d can be taken as 1 or 2) such that

$$\sum_{i=1}^{t} a_{id} \leq \frac{t}{2} - 2014.$$

Let $b_s = \sum_{i=1}^{s} a_{id} - \dfrac{s}{2}$, $s = 1, 2, \ldots, t$, then $b_1 = a_d - \dfrac{1}{2} \geq -\dfrac{1}{2}$, and $b_t \leq -2014$. Let l be the smallest positive integer satisfying $b_l \leq -2014$, then $b_{l-1} \geq -2014 + \dfrac{1}{2}$. Since $b_l - b_{l-1} = a_{ld} - \dfrac{1}{2} \geq -\dfrac{1}{2}$, we have $b_l \geq b_{l-1} - \dfrac{1}{2} \geq -2014$, thus $b_l = -2014$, i.e.

$$\sum_{i=1}^{l} a_{id} = \frac{l}{2} - 2014.$$

Since $\sum_{i=1}^{1} a_{id}$ is an integer, l must be even. Write $l = 2k$, thus k is a positive integer, and

$$\sum_{i=1}^{2k} a_{id} = k - 2014. \qquad \square$$

China Mathematical Olympiad

First Day

December 16, 2015

8:00–12:30

1. Let $a_1, a_2, \ldots, a_{31}, b_1, b_2, \ldots, b_{31}$ be positive integers satisfying

(i) $a_1 < a_2 < \cdots < a_{31} \leq 2015$, $b_1 < b_2 < \cdots < b_{31} \leq 2015$; and

(ii) $a_1 + a_2 + \cdots + a_{31} = b_1 + b_2 + \cdots + b_{31}$.

Find the maximum value of the sum $S = |a_1 - b_1| + |a_2 - b_2| + \cdots + |a_{31} - b_{31}|$.

Solution Let $A = \{m : a_m > b_m, \ 1 \leq m \leq 31\}$ and $B = \{n : a_n < b_n, \ 1 \leq n \leq 31\}$. Write

$$S_1 = \sum_{m \in A} (a_m - b_m), \quad \text{and} \quad S_2 = \sum_{n \in B} (b_n - a_n),$$

then $S = S_1 + S_2$. It follows by (ii) that

$$S_1 - S_2 = \sum_{m \in A \cup B} (a_m - b_m) = 0,$$

hence $S_1 = S_2 = \dfrac{S}{2}$.

If $A = \emptyset$, then $S = 2S_1 = 0$. In what follows, we assume $A \neq \emptyset$, and $B \neq \emptyset$. Now $|A|, |B|$ are positive integers with $|A| + |B| \leq 31$. Put

$$u = a_k - b_k = \max_{m \in A}(a_m - b_m), \quad \text{and} \quad v = b_l - a_l = \max_{m \in B}(b_n - a_n).$$

45

We claim that $u + v \leq 1984$. Without loss of generality, we may assume $1 \leq k < l \leq 31$. Then

$$u + v = (a_k - b_k) + (b_l - a_l) = b_{31} - (b_{31} - b_l) - b_k - (a_l - a_k).$$

It follows by (i) that $b_{31} \leq 2015$, $b_{31} - b_l \geq 31 - l$, $b_k \geq k$ and $a_l - a_k \geq l - k$. Thus

$$u + v \leq 2015 - (31 - l) - k - (l - k) = 1984.$$

It is clear that $S_1 \leq u|A|$ and $S_2 \leq v|B|$. Therefore

$$1984 \geq u + v \geq \frac{S_1}{|A|} + \frac{S_2}{|B|} \geq \frac{S_1}{|A|} + \frac{S_2}{31 - |A|}$$

$$= \frac{S}{2} \cdot \frac{31}{|A|(31 - |A|)} \geq \frac{31S}{2 \times 15 \times 16},$$

hence $S \leq \dfrac{2 \times 15 \times 16 \times 1984}{31} = 30720.$

It is easily verified that

$$(a_1, a_2, \ldots, a_{16}, a_{17}, a_{18}, \ldots, a_{31}) = (1, 2, \ldots, 16, 2001, 2002, \ldots, 2015),$$

and

$$(b_1, b_2, \ldots, b_{31}) = (961, 962, \ldots, 991)$$

satisfy both conditions (i) and (ii), and S achieves the maximum value 30720.

The required maximum value of S is 30720.   □

2   As shown in Figure 2.1, $ABCD$ is a convex quadrilateral. Points K, L, M, N lie on the sides AB, BC, CD, DA respectively, such that

$$\frac{AK}{KB} = \frac{DA}{BC}, \quad \frac{BL}{LC} = \frac{AB}{CD}, \quad \frac{CM}{MD} = \frac{BC}{DA}, \quad \frac{DN}{NA} = \frac{CD}{AB}.$$

The extensions of AB and DC meet at point E, and the extensions of AD and BC meet at point F. The inscribed circle of triangle AEF touches the sides AE, AF at points S, T respectively; the inscribed circle of triangle CEF touches the sides CE, CF at points U, V respectively.

Prove that if points K, L, M, N are concyclic, then points S, T, U, V are concyclic.

Proof. Let $AB = a$, $BC = b$, $CD = c$ and $DA = d$. It follows from the assumption that

$$AK = \frac{ad}{b+d}, \quad BK = \frac{ab}{b+d}, \quad BL = \frac{ab}{a+c}, \quad CL = \frac{bc}{a+c},$$

$$CM = \frac{bc}{b+d}, \quad DM = \frac{cd}{b+d}, \quad DN = \frac{cd}{a+c}, \quad AN = \frac{ad}{a+c}.$$

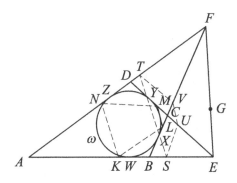

Fig. 2.1

If $a + c > b + d$, then $AK > AN$, and $\angle AKN < \angle KNA$. Similarly, we have

$$\angle BKL < \angle KLB, \quad \angle CML < \angle MLC, \quad \text{and} \quad \angle DMN < \angle MND.$$

It follows that

$$2\pi - \angle AKN - \angle BKL - \angle CML - \angle DMN > 2\pi$$

$$-\angle KNA - \angle KLB - \angle MLC - \angle MND,$$

i.e. $\angle NML + \angle NKL > \angle MNK + \angle MLK$, which contradicts with K, L, M, N being concyclic. Hence $a + c \leq b + d$. Similarly one proves that $a + c \geq b + d$, thus $a + c = b + d$, which implies that $ABCD$ has an inscribed circle ω.

Denote the points that ω touches the sides AB, BC, CD, DA by W, X, Y, Z, respectively. We have

$$AE - AF = WE - ZF = EY - FX = EC - CF.$$

Suppose the inscribed circles of triangle AEF and CEF touch EF at points G and H respectively, then

$$2(FG - FH) = (EF + AF - AE) - (EF + CF - CE)$$
$$= (AF - AE) - (CF - CE) = 0,$$

hence G and H coincide. Since $ES = EG = EU$ and $FT = FG = FV$, we have

$$\angle EUS = \frac{\pi - \angle UES}{2} = \frac{\angle A + \angle ADC}{2},$$

$$\angle FTV = \frac{\pi - \angle TFV}{2} = \frac{\angle A + \angle ABC}{2}$$

Noting that $\angle ATS = \frac{1}{2}(\pi - \angle A)$ and $\angle CUV = \frac{1}{2}(\pi - \angle BCD)$, thus

$$\angle VTS + \angle VUS = (\pi - \angle FTV - \angle ATS) + (\angle CUV + \pi - \angle EUS)$$
$$= \left(\pi - \frac{\angle A + \angle ABC}{2} - \frac{\pi - \angle A}{2}\right)$$
$$+ \left(\frac{\pi - \angle BCD}{2} + \pi - \frac{\angle A + \angle ADC}{2}\right)$$
$$= \pi,$$

hence S, T, U, V are concyclic. This completes the proof. $\qquad\square$

3 Let p be an odd prime, and $a_1, a_2, \ldots, a_p$ be integers. Prove that the following two statements are equivalent:

(I) There exists a polynomial $f(x)$ with integer coefficients and of degree less than or equal to $\frac{p-1}{2}$, such that $f(i) \equiv a_i \pmod{p}$ holds for every positive integer $i \le p$.

(II) For each positive integer $d \le \frac{p-1}{2}$, the congruence equality

$$\sum_{i=1}^{p}(a_{i+d} - a_i)^2 \equiv 0 \pmod{p}$$

holds, where the indices are considered modulo p, that is, $a_{p+n} = a_n$.

Proof. We need the following results for this problem.

(1) The difference of f is defined as $\Delta f = \Delta f(x) = f(x+1) - f(x)$, higher order differences are defined recursively as

$$\Delta^0 f = f, \quad \Delta^1 f = \Delta f, \quad \Delta^n f = \Delta(\Delta^{n-1}f), \quad n = 2, 3, \ldots.$$

If $\deg(f) \geq 1$, then $\deg(\Delta f) = \deg(f) - 1$ and the leading coefficient of Δf is the leading coefficient of f multiplied by $\deg(f)$. If $\deg(f) = 0$, then $\Delta(f)$ is the zero polynomial.

(2) For any positive integer n, we have

$$f(x + n) = \sum_{i=0}^{n} C_n^i \Delta^i f(x).$$

This is the inversion formula of the standard difference formula

$$\Delta^n f(x) = \sum_{i=0}^{n} (-1)^i C_n^i f(x + n - i).$$

It can also be verified directly by induction on n.

(3) For any $d \in \mathbb{Z}$, put

$$T_d = \sum_{x=1}^{p} (f(x + d)) - f(x))^2.$$

Thus $T_0 = 0$, and $T_{p-d} \equiv T_d \equiv T_{p+d} \pmod{p}$. For any $i > 0$, put

$$S_i = \sum_{x=1}^{p} \Delta^i f(x) \cdot f(x).$$

We show that T_d can be represented as a linear combination of $S_1, S_2, \ldots$ modulo p. Indeed for $d > 0$,

$$T_d = \sum_{x=1}^{p} f^2(x + d) + \sum_{x=1}^{p} f^2(x) - 2 \sum_{x=1}^{p} f(x + d) f(x)$$

$$\equiv 2 \sum_{i=1}^{p} f^2(x) - 2 \sum_{x=1}^{p} f(x + d) f(x)$$

$$= -2 \sum_{x=1}^{p} (f(x + d) - f(x)) \cdot f(x)$$

$$= -2 \sum_{x=1}^{p} \left(\sum_{i=0}^{d} C_d^i \Delta^i f(x) - f(x) \right) \cdot f(x) = -2 \sum_{x=1}^{p} \sum_{i=1}^{d} C_d^i \Delta^i f(x) \cdot f(x)$$

$$= -2 \sum_{i=1}^{d} C_d^i \sum_{x=1}^{p} \Delta^i f(x) \cdot f(x)$$

$$= -2 \sum_{i=1}^{d} C_d^i S_i \pmod{p}.$$

(4) We will frequently use the following congruence equality:

$$\sum_{x=1}^{p} x^k \equiv \begin{cases} 0 \ (\text{mod } p), & k = 0, 1, \ldots, p - 2, \\ -1 \ (\text{mod } p), & k = p - 1. \end{cases}$$

(5) Let $g(x) = B_{p-1}x^{p-1} + \cdots + B_1 x + B_0$ be an integral polynomial. It follows by (4) that

$$\sum_{x=1}^{p} g(x) \equiv -B_{p-1} \ (\text{mod } p).$$

In particular, $\sum_{x=1}^{p} g(x) \equiv 0 \ (\text{mod } p)$ if $\deg(g) \leq p - 2$.

We now proceed to prove the original problem. First assume that statement (I) holds and let $f(x)$ be a polynomial as in (I). If f is a constant polynomial, then clearly $T_d = 0$ for all d.

Assume that $1 \leq \deg(f) \leq \dfrac{p-1}{2}$. Noting that

$$\deg(\Delta^i f \cdot f) \leq 2 \deg(f) - 1 \leq p - 2,$$

it follows from (5) that

$$S_i = \sum_{x=1}^{p} \Delta^i f(x) \cdot f(x) \equiv 0 \ (\text{mod } p),$$

thus $T_d \equiv 0 \ (\text{mod } p)$ by (3) for all $d \leq \dfrac{p-1}{2}$. Therefore statement (II) holds.

Now assume that statement (II) holds. For each $i \in \{1, 2, \ldots, p\}$, choose an integer λ_i satisfying

$$\lambda_i \cdot \prod_{1 \leq j \leq p, j \neq i} (i - j) \equiv p \ (\text{mod } p),$$

and define

$$f(x) \equiv \sum_{i=1}^{p} \left(a_i \lambda_i \prod_{1 \leq j \leq p, j \neq i} (x - j) \right) \ (\text{mod } p),$$

where the leading coefficient of f is not divisible by p unless f is the zero polynomial. Clearly f is an integral polynomial with $\deg(f) \leq p - 1$ and

$f(i) \equiv a_i \pmod{p}$ for each $i = 1, 2, \ldots, p$. Assume that f is not the zero polynomial, put

$$f(x) = \sum_{i=0}^{m} B_i x^i \pmod{p},$$

$B_m \equiv 0 \pmod{p}$. We shall show that $m \le \dfrac{p-1}{2}$. Assume on the contrary that $m > \dfrac{p-1}{2}$. We have

$$T_d = \sum_{x=1}^{p} (f(x+d) - f(x))^2 \equiv \sum_{i=1}^{p} (a_{i+d} - a_i)^2 \equiv 0 \pmod{p},$$

for $d = 1, 2, \ldots, \dfrac{p-1}{2}$. The above congruence holds in fact for all positive integers d since $T_{p-d} = \equiv T_d \equiv T_{p+d} \pmod{p}$. By the result in (3), we get inductively that $S_i \equiv 0 \pmod{p}$ for all $i \ge 1$.

On the other hand, let $k = 2m - (p-1)$, then $0 < k \le m$. By (1) we know that $\deg(\Delta^k f) = m - k$ with leading coefficient $m(m-1) \cdots (m-k+1)B_m$. Hence $\deg(\Delta^k f \cdot f) = m - k + m = p - 1$, with leading coefficient $m(m-1) \cdots (m-k+1)B_m^2$, and it follows from (5) that

$$S_k = \sum_{x=1}^{p} \Delta^k f(x) \cdot f(x) \equiv -m(m-1) \cdots (m-k+1)B_m^2 \equiv 0 \pmod{p},$$

contradicting the fact that $S_i \equiv 0 \pmod{p}$ for all $i \ge 1$. Thus $m \le \dfrac{p-1}{2}$, and $\deg(f) \le \dfrac{p-1}{2}$.

Statement (I) is true.

This completes the proof of the equivalence of the two statements. $\square$

Second Day

December 17, 2015

8:00–12:30

4 Let $n \geq 3$ be an integer, and k be the number of primes not exceeding n. Let A be a subset of $\{2, 3, \ldots, n\}$ whose cardinality is less than k, satisfying that any element of A is not a multiple of any other element of A. Prove that there exists a k-element subset B of $\{2, 3, \ldots, n\}$, such that any element of B is not a multiple of any other element of B, and B contains A.

Proof. For an integer $m > 1$, let

$$m = p_1^{\alpha_1} \cdots p_l^{\alpha_l}$$

be the standard factorization and define $f(m) = \max\{p_1^{\alpha_1}, \ldots, p_l^{\alpha_l}\}$. Since $|A| < k$, there exists a prime $p \leq n$ such that p does not divide any $f(a)$, $a \in A$. Let α be the largest positive integer satisfying $p^\alpha \leq n$, that is $p^\alpha \leq n < p^{\alpha+1}$. We show that $a \nmid p^\alpha$ and $p^\alpha \nmid a$ for all $a \in A$.

If $a \mid p^\alpha$, then $p \mid f(a) = a$, contradicting the choice of p, hence $a \nmid p^\alpha$.

If $p^\alpha \mid a$, then $f(a) = q^\beta$ where q is a prime number distinct from p since $p \nmid f(a)$. By definition, $q^\beta > p^\alpha \geq p$, hence

$$f(a) \geq q^\beta p^\alpha \geq p \cdot p^\alpha = p^{\alpha+1} > n,$$

a contradiction.

Note in particular that $p^\alpha \notin A$. If we join p^α in A, then A still satisfies the property that any element is not a multiple of any other element. Repeat this process until we get a k-element subset B of $\{2, 3, \ldots, n\}$, which has the required property.   □

5 Let $ABCD$ be an arbitrary convex quadrilateral in the plane. Prove that there exists a square $A'B'C'D'$ (the vertices may be ordered clockwisely or counterclockwisely as you want), such that $A' \neq A$, $B' \neq B$, $C' \neq C$, $D' \neq D$, and the lines AA', BB', CC', DD' pass through a common point.

Proof. We distinguish two cases.

Case 1 : *ABCD* is a rectangle as shown in Figure 5.1.

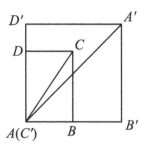

Fig. 5.1

Choose a point A' on the angle bisector of $\angle BAD$ such that the orthogonal projection points of A' onto the lines AB and AD (denoted by B' and D' respectively) are distinct from B and D. Let $C' = A \neq C$, thus $A'B'C'D'$ is a square and the lines AA', BB', CC', DD' all pass through A.

Case 2: *ABCD* is not a rectangle. We may assume without loss of generality that $\angle BAD$ is an acute angle as shown in Figure 5.2.

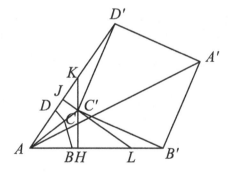

Fig. 5.2

Pick a point C' on the extension of AC such that the orthogonal projection points of C' onto the lines AB and AD (denoted by H and J respectively) are on the extensions of AB and AD. Since $\angle BAD$ is acute, the line HC' meets the extension of AJ at point K and the line JC' meets the extension of AH at point L. We choose points B' and D' on the extensions of AL and AK respectively such that $LB' = KC'$ and $KD' = LC'$. Note that $\angle B'LC' = 90° + \angle LAK = \angle C'KD'$, we have that $\triangle B'LC' \cong \triangle C'KD'$. Thus $B'C' = C'D'$, and

$$\angle B'C'D' = 180° - \angle KC'D' - \angle B'C'H$$

$$= 180° - \angle LB'C' - (90° - \angle LB'C') = 90°.$$

Let A' be the fourth point to make $A'B'C'D'$ a square, then A lies outside $AHCJ$, and AA', BB', CC', DD' all pass through A. This completes the proof. □

6 100 players participate in a tournament. Any two players x and y play exactly once and there is no draw in this game. We use $x \to y$ to mean that x beats y. The tournament is called *friendly* if for each pair of players x, y, there exists a sequence of players $u_1, u_2, \ldots, u_k$ $(k \geq 2)$, such that $x = u_1 \to u_2 \to \cdots \to u_k = y$.

(a) For any friendly tournament, prove that there exists a positive integer m, such that for each pair of players x, y, there exists a sequence of players $z_1, z_2, \ldots, z_m$ of length m (repetition is allowed in $z_1, z_2, \ldots, z_m$), satisfying $x = z_1 \to z_2 \to \cdots \to z_m = y$.

(b) For a friendly tournament T, let $m(T)$ denote the minimal value of m as stated in (a). Find the smallest possible value of $m(T)$.

Solution   (a) The tournament in the problem is exactly the tournament in graph theory, that is a complete digraph. Friendly means strongly connected. We need the following lemmas.

Lemma 1   *In any strongly connected tournament with $n \geq 3$ vertices, there is a directed cycle of length n.*

Lemma 2   *In any strongly connected tournament with $n \geq 4$ vertices, there is a directed cycle of length $n - 1$.*

Lemma 3   *Let a, b be coprime positive integers, then every positive integer $m \geq ab - a - b + 1$ can be represented as $m = ax + by$ for some non-negative integers x, y.*

Lemma 3 is a well-known result in elementary number theory whose proof will be omitted. Lemma 1 and 2 will be proved in the end of the solution.

For a strongly connected tournament with $n \geq 4$ vertices, we show that $m = n^2$ works for (a). By lemma 1 and 2, there are directed cycles C_1 and C_2 of length n and $n - 1$ respectively. Let x, y be any two players and fix a path P from x to y with $s \leq n - 1$ edges.

Since $m - s - 1 \geq n^2 - n = n(n - 1)$, it follows from lemma 3 that there exist nonnegative integers a, b satisfying $na + (n - 1)b = m - s - 1$.

If y lies on C_2, we start from x, go along C_1 for a rounds, go along P, then go along C_2 for b rounds. This gives a directed path from x to y with $an + b(n - 1) + s = m - 1$ edges, which yields the required sequence of m players.

If x lies on C_2, we start from x, go along C_2 for b rounds, go along P, then go along C_1 for a rounds which is a directed path from x to y with $m - 1$ edges. This proves part (a).

(b) Let T be a strongly connected tournament with 100 vertices. Let x, y be two players such that $x \to y$, then a directed path from y to x has at least 3 vertices, hence $m(T) \geq 3$. We now construct a strongly connected tournament T_0 with $m(T_0) = 3$.

Denote the vertices by $v_1, v_2, \ldots, v_{100}$. Let

$$A = \{i \in \mathbb{Z} : 1 \leq i \leq 49, \ i \neq 3\} \cup \{97\}.$$

It is easily verified that

$$A + A = \{a + b : a, b \in A\} = \{2, 3, \ldots, 98\} \cup \{101, 102, \ldots, 46\}.$$

We see that A has the following two properties:

(i) For $i = 1, 2, \ldots, 49$, exactly one of i and $100 - i$ belongs to A.
(ii) For any integer $k \not\equiv 0 \pmod{100}$, there exist $a, b \in A$ such that $a + b \equiv k \pmod{100}$.

Define the edges of T_0 as follows: $v_k \to v_{k+50}$ for $k = 1, 2, \ldots, 50$. For $1 \leq i, j \leq 100$, $i - j \not\equiv 0, 50 \pmod{100}$, we define $v_i \to v_j$ if $j - i \equiv a \pmod{100}$ for some $a \in A$. By (i), exactly one of $v_i \to v_j$ and $v_j \to v_i$ holds, hence the edge between v_i, v_j is well-defined. For any two vertices v_i, v_j, $i \neq j$, there exists $a, b \in A$ such that $a + b \equiv j - i \pmod{100}$ by (ii), hence $v_i \to v_{i+a} \to v_{i+a+b} = v_j$ (here the indices are considered modulo 100). Thus $m(T_0) = 3$.

The smallest possible value of $m(T)$ is 3.

Proof of lemma 1. Let T be a strongly connected tournament with $n \geq 3$ vertices. Since there are a path from x to y and a path from y to x, there exists a directed cycle in T. Let C be the largest simple cycle in T. We shall show that C is a Hamiltonian cycle.

Let C be $v_1 \rightarrow v_2 \rightarrow \cdots \rightarrow v_p \rightarrow v_1$. Suppose on the contrary that $p \leq n - 1$. Let u be a vertex not on C. We consider several cases.

Case 1: $u \rightarrow v_i$ for all $1 \leq i \leq p$. Consider the shorted directed path from $\{v_1, \ldots, v_p\}$ to u, say $v_p \rightarrow z_1 \rightarrow \cdots \rightarrow z_k \rightarrow u$, $k \geq 1$, and $z_1, \ldots, z_k \notin \{v_1, \ldots, v_p\}$. The simple cycle

$$v_1 \rightarrow \cdots \rightarrow v_p \rightarrow z_1 \rightarrow \cdots \rightarrow z_k \rightarrow u \rightarrow v_1$$

has $p + k + 1 > p$ edges. Contradiction to the maximality of C.

Case 2: $v_i \rightarrow u$ for all $1 \leq i \leq p$. Reverse the direction of each edge of T, and it is reduced to case 1.

Case 3: $v_i \rightarrow u \rightarrow v_j$ for some i, j. Without loss of generality, we may assume $u \rightarrow v_p$. Let k be the largest integer i such that $v_i \rightarrow u$, then $k < p$ and $v_k \rightarrow u \rightarrow v_{k+1}$. Hence the following simple cycle

$$v_1 \rightarrow \cdots \rightarrow v_k \rightarrow u \rightarrow v_{k+1} \rightarrow \cdots \rightarrow v_p \rightarrow v_1$$

has $p + 1 > p$ edges, a contradiction.

This proves lemma 1.

Proof of lemma 2. Let T be a strongly connected tournament with $n \geq 4$ vertices. By lemma 1, there exists a Hamiltonian cycle

$$x_1 \rightarrow x_1 \rightarrow \cdots \rightarrow x_n \rightarrow x_1.$$

One finds a proper simple cycle (a simple cycle with less than n edges) by the edge between x_1 and x_3 no matter $x_1 \rightarrow x_3$ or $x_3 \rightarrow x_1$.

Let H be a proper strongly connected sub-tournament of T with vertices $v_1, v_2, \ldots, v_p$ such that p is largest possible. We show that $p = n - 1$.

Clearly $p \geq 3$. Assume on the contrary that $p < n - 1$. Let u be a vertex not in H. Let

$$U_1 = \{v \in T \mid v \rightarrow v_1\}, \quad U_2 = \{v \in T \mid v_1 \rightarrow v\}.$$

If $u \in U_1$, then $u \rightarrow v_i$ for all $1 \leq i \leq p$, otherwise $u, v_1, \ldots, v_p$ induce a larger strongly connected proper subgraph of T. Similarly, $u \in U_2$ implies $v_i \rightarrow u$ for all $1 \leq i \leq p$.

Since T is strongly connected, $U_1, U_2 \neq \emptyset$, and there exist $u_1 \in U_1$ and $u_2 \in U_2$ such that $u_2 \to u_1$. Thus $u_1, u_2, v_1, v_2, \ldots, v_{p-1}$ induce a strongly connected subgraph with $p+1$ vertices, contradicting the maximality of H. Therefore H has $n-1$ vertices, and a simple cycle with $n-1$ edges in T is found by lemma 1 as a Hamiltonian cycle in H. This proves lemma 2. $\quad\square$

China National Team Selection Test

2015

First Day

March 23, 2015

8:00–12:30

1. In an isosceles triangle ABC, $AB = AC > BC$, a point D is inside $\triangle ABC$ and $DA = DB + DC$. The perpendicular bisector of AB intersects the exterior angle bisector of $\angle ADB$ at the point P; the perpendicular bisector of AC intersects the exterior angle bisector of $\angle ADC$ at the point Q. Prove that B, C, P, and Q are concyclic. (posed by He Yijie)

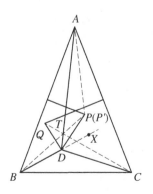

Fig. 1.1

Proof. We first prove that A, B, D, and P are concyclic. Indeed, let P' be the midpoint of the arc $\overset{\frown}{ADB}$, P' is on the perpendicular bisector of AB. Extend BD to any point X. Since $P'A = P'B$ and $A, B, D,$ and P' are concyclic, we have $\angle P'DA = \angle P'BA = \angle P'AB = \angle P'DX$, so P' is also on the exterior angle bisector of $\angle ADB$, thus $P' = P$, and $A, B, D,$ and P are concyclic.

By Ptolemy's theorem, $AB \cdot DP + BD \cdot AP = AD \cdot BP$; and note that $PA = PB$, and $AD = BD + CD$, we get

$$AB \cdot DP = AD \cdot BP - BD \cdot AP = AP \cdot (AD - BD) = AP \cdot CD;$$

so $\dfrac{AP}{DP} = \dfrac{AB}{CD}$.

Let T be the intersection of BP and AD, note that $\angle BAP + \angle BDP = 180°$, we have

$$\frac{AT}{TD} = \frac{S_{\triangle ABP}}{S_{\triangle DBP}} = \frac{\dfrac{1}{2}AB \cdot AP \cdot \sin \angle BAP}{\dfrac{1}{2}DB \cdot DP \cdot \sin \angle BDP}$$

$$= \frac{AB}{DB} \cdot \frac{AP}{DP} = \frac{AB}{DB} \cdot \frac{AB}{CD} = \frac{AB^2}{BD \cdot CD}$$

By the same reason, $A, C, D,$ and Q are concyclic; and let T' be the intersection of CQ and AD, we have

$$\frac{AT'}{T'D} = \frac{AC^2}{BD \cdot CD}.$$

Since $AB = AC$, so $\dfrac{AT'}{T'D} = \dfrac{AT}{TD}$. Thus $T = T'$.

By the power of a point theorem, $TB \cdot TP = TA \cdot TD = TC \cdot TQ$, so $B, C, P,$ and Q are concyclic. $\qquad\square$

2 X is a non-empty finite set, and $A_1, A_2, \ldots, A_k$ are k subsets of X satisfying:

(i) $|A_i| \le 3$, $i = 1, 2, \ldots, k$; and

(ii) Any element of X is contained in at least 4 of $A_1, A_2, \ldots, A_k$.

Prove that we can pick $\left\lfloor \dfrac{3}{7}k \right\rfloor$ sets from $A_1, A_2, \ldots, A_k$, so that the union of the picked sets is X. (posed by Xiong Bin)

Proof (First proof). We first pick a maximal pairwise disjoint sub-collection of 3-element subsets among $A_1, A_2, \ldots, A_k$. Without loss of generality, it is $A_1, A_2, \ldots, A_x$. For each i, where $x < i \le k$, let

$$B_i = A_i \setminus (\cup_{j=1}^{x} A_j),$$

so $|B_i| \le 2$ — otherwise A_i is of size 3 and disjoint from $A_1, A_2, \ldots, A_x$, contradicts the maximality of x. Let

$$X_1 = X \setminus (\cup_{j=1}^{x} A_j).$$

By (ii), each element of X_1 belongs to at least 4 of $B_{x+1}, \ldots, B_k$.

Next we pick a maximal pairwise disjoint sub-collection of 2-element subsets among $B_{x+1}, \ldots, B_k$. Without loss of generality, we picked y sets $B_{x+1}, \ldots, B_{x+y}$. For each i, where $x + y < i \le k$, let

$$C_i = B_i \setminus (\cup_{j=x+1}^{x+y} B_j);$$

then $|C_i| \le 1$. Let

$$X_2 = X_1 \setminus (\cup_{j=x+1}^{x+y} B_j);$$

each element of X_2 belongs to at least 4 of $C_{x+y+1}, \ldots, C_k$. Finally we pick z sets from $C_{x+y+1}, \ldots, C_k$ so that their union is X_2. Without loss of generality, they are $C_{x+y+1}, \ldots, C_{x+y+z}$.

Consider the sets $A_1, A_2, \ldots, A_{x+y+z}$. We have

$$\cup_{i=1}^{x+y+z} A_i = (\cup_{i=1}^{x} A_i) \cup (\cup_{i=x+1}^{x+y} B_i) \cup (\cup_{i=x+y+1}^{x+y+z} C_i)$$

$$= (X \setminus X_1) \cup (X_1 \setminus X_2) \cup X_2 = X. \qquad \textcircled{1}$$

Now we prove that $x + y + z < \dfrac{3}{7}k$.

Clearly, $|X_1| = |X| - 3x$, $|X_2| = |X_1| - 2y = |X| - 3x - 2y$. By (i) and (ii), it is easy to see

$$4|X| \le \sum_{i=1}^{k} |A_i| \le 3k.$$

Similarly, by the properties of B_i's and C_j's,

$$4(|X| - 3x) = 4|X_1| \leq \sum_{i=x+1}^{k} |B_i| \leq 2(k - x),$$

$$4(|X| - 3x - 2y) = 4|X_2| \leq \sum_{i=x+y+1}^{k} |C_i| \leq k - x - y.$$

From the three inequalities above we get

$$|X| \leq \frac{3k}{4}, \quad 5x \geq 2|X| - k, \quad 11x + 7y \geq 4|X| - k.$$

So we get

$$x + y + z = x + y + |X| - 3x - 2y = |X| - 2x - y$$

$$= |X| - \frac{3}{35} \cdot 5x - \frac{1}{7}(11x + 7y)$$

$$\leq |X| - \frac{3}{35} \cdot (2|X| - k) - \frac{1}{7}(4|X| - k)$$

$$= \frac{9}{35}|X| + \frac{8}{35}k \leq \frac{9}{35} \cdot \frac{3}{4}k + \frac{8}{35}k$$

$$= \frac{59}{140}k < \frac{3}{7}k.$$

So, $x + y + z \leq \left\lfloor \frac{3}{7}k \right\rfloor$, and by ①, the union of $A_1, A_2, \ldots, A_{x+y+z}$ is X.

Proof (based on the solution from Yu Chenjie in the test). We may assume that each element of X belongs to exactly 4 of the sets $A_1, A_2, \ldots, A_k$ — otherwise we may remove this element from some of the sets without affecting the problem. Now consider a minimum, with respect to the number of sets picked, sub-collection of $A_1, A_2, \ldots, A_k$ whose union is X. Without loss of generality, it is $A_1, A_2, \ldots, A_m$. For each integer s, where $1 \leq s \leq m$, there exists an element $x_s \in A_s$ so that, among $A_1, A_2, \ldots, A_m$, x_s only belongs to A_s — otherwise, remove A_s from $A_1, \ldots, A_m$, the union of the resulting sub-collection is still X, contradicts the minimality of m. Clearly $x_1, x_2, \ldots, x_s$ are distinct.

Let $\mathcal{U} = \{A_1, A_2, \ldots, A_m\}$ and $\mathcal{V} = \{A_{m+1}, \ldots, A_k\}$. Construct a bipartite graph $G = G(\mathcal{U}, \mathcal{V})$ as follows: for each $A_s \in \mathcal{U}$ and $A_t \in \mathcal{V}$, add an edge between A_s and A_t if $x_s \in A_t$. For each $A_s \in \mathcal{U}$, x_s belongs to 4 sets, exactly 3 of them are in $\mathcal{V}$, so $\deg_G(A_s) = 3$. And by condition (i),

$\deg_G(A_t) \leq 3$ for each $A_t \in \mathcal{V}$. Let S be the set of all pairs $(\{A_{s_1}, A_{s_2}\}, A_t)$, where A_{s_1} and A_{s_2} are two distinct elements in $\mathcal{U}$, $A_t \in \mathcal{V}$ and is adjacent to both A_{s_1} and A_{s_2}. Suppose there are a elements in X each of which belongs to exactly one set in $\mathcal{U}$, these clearly include $x_1, x_2, \ldots, x_m$, we denote them by $x_1, x_2, \ldots, x_a$; and let $y_1, y_2, \ldots, y_b$ be those elements belonging to exactly 2 sets in $\mathcal{U}$.

For any $(\{A_{s_1}, A_{s_2}\}, A_t) \in S$, there exists an element $z \in A_{s_1} \cup A_{s_2}$ such that $z \neq x_{s_1}$, $z \neq x_{s_2}$, and z does not belong to any set in $\mathcal{U}$ other than A_{s_1} and A_{s_2} — otherwise, we can remove A_{s_1} and A_{s_2} from $\mathcal{U}$ and add A_t, contradicts the minimality of m. Since z belongs to at most two sets in $\mathcal{U}$, and it is not one of $x_1, \ldots, x_m$, so $z \in Z = \{x_{m+1}, \ldots, x_a, y_1, \ldots, y_b\}$. For each $(\{A_{s_1}, A_{s_2}\}, A_t) \in S$, pick such a $z \in Z$, we get a mapping $f : S \to Z$.

Next we construct another bipartite graph $H = H(\mathcal{V}, Z)$: For $A_t \in \mathcal{V}$, $z \in Z$, add an edge between them if there exists $(\{A_{s_1}, A_{s_2}\}, A_t) \in S$ such that $f(\{A_{s_1}, A_{s_2}\}, A_t) = z$.

On one hand, we show that $\deg_H z \leq 3$ for each $z \in Z$: For each x_i, $m < i \leq a$, suppose A_s, $1 \leq s \leq m$, is the unique set in $\mathcal{U}$ that contains x_i. Suppose x_i and A_t are adjacent in H, then A_s and A_t are adjacent in G; since $\deg_G(A_s) = 3$, x_i is adjacent in H to at most 3 A_t's, i.e., $\deg_H(x_i) \leq 3$. For each y_j, $1 \leq j \leq b$, suppose A_{s_1}, A_{s_2}, $1 \leq s_1 < s_2 \leq m$, are the two sets in $\mathcal{U}$ that contain y_j. Suppose y_j is adjacent to A_t in H, then A_{s_1} and A_{s_2} are both adjacent to A_t in G; yet $\deg_G(A_{s_1}) = \deg_G(A_{s_2}) = 3$, so there are at most 3 such A_t's.

On the other hand, for any $A_t \in \mathcal{V}$, when $\deg_G(A_t) = 2$, denoting A_{s_1} and A_{s_2} its neighbours in G, A_t is adjacent in H to $f(\{A_{s_1}, A_{s_2}\}, A_t)$, so $\deg_H(A_t) \geq 1$; when $\deg_G(A_t) = 3$, denoting A_{s_1}, A_{s_2}, and A_{s_3} its neighbours in G, without loss of generality, $f(\{A_{s_1}, A_{s_2}\}, A_t) \in A_{s_1}$, so $f(\{A_{s_2}, A_{s_3}\}, A_t) \neq f(\{A_{s_1}, A_{s_2}\}, A_t)$, and $\deg_H(A_t) \geq 2$. Let α and β be the numbers of sets $A_t \in \mathcal{V}$ with $\deg_G(A_t) = 2$ and $\deg_G(A_t) = 3$, respectively. Calculate the number of edges in G by counting the sum of degrees on both sides, we get

$$3m = \sum_{A_s \in \mathcal{U}} \deg_G(A_s) = \sum_{A_t \in \mathcal{V}} \deg_G(A_t) \leq 2\alpha + 3\beta + (k - m - \alpha - \beta).$$

Count the sum of degrees on both sides in H, we get

$$\alpha + 2\beta \leq \sum_{A_t \in \mathcal{V}} \deg_H(A_t) = \sum_{z \in Z} \deg_H(z) \leq 3(a - m + b).$$

Combining the two inequalities above, we get

$$4m - k \leq 3(a - m + b).$$

Also each x_i, where $1 \leq i \leq a$, belongs to exactly 3 sets in $\mathcal{V}$; each y_j, where $1 \leq j \leq b$, belongs to exactly 2 sets in $\mathcal{V}$, so

$$3(k - m) \geq \sum_{i=m+1}^{k} |A_i| \geq 3a + 2b.$$

Note that $a \geq m$, we have

$$3(k - m) \geq 3a + 2b = \frac{2}{3} \cdot 3(a - m + b) + 2m + a$$

$$\geq \frac{2}{3}(4m - k) + 2m + m,$$

which implies $m \leq \dfrac{11}{26}k$, so $m \leq \left\lfloor \dfrac{11}{26}k \right\rfloor \leq \left\lfloor \dfrac{3}{7}k \right\rfloor$ $\square$

③ Let a and b be two positive integers whose greatest common divisor has at least two different prime factors. Let

$$S = \{n \in \mathbb{N}^* : n \equiv a \pmod{b}\}.$$

Call an element $x \in S$ *irreducible* if it can not be expressed as the product of two or more (not necessarily distinct) elements of S. Prove that there exists a positive integer t such that each elements in S can be expressed as the product of up to t irreducible elements in S. (posed by Qu Zhenhua)

Proof. First note that every element in S can be expressed, not necessary uniquely, as the product of irreducible elements. Moreover, if $x \in S$ is the product of m elements in S, where $m \geq 2$, take modulo b, we get $a^m \equiv a \pmod{b}$. So, when there does not exist $m \geq 2$ where $a^m \equiv a \pmod{b}$, each element in S is irreducible and we are done.

Otherwise, pick the smallest positive integer $m_0 \geq 2$ so that $a^{m_0} \equiv a \pmod{b}$. We prove,

$$a^n \equiv a \pmod{b} \Leftrightarrow n \equiv 1 \pmod{m_0 - 1}. \qquad ①$$

First assume that $a^n \equiv a \pmod{b}$. Suppose $n - 1 = (m_0 - 1)q + r$, $0 \leq r < m_0 - 1$. When $q > 0$, we have $n = (m_0 - 1)(q - 1) + r + m_0$, and

$$a \equiv a^n \equiv a^{(m_0-1)(q-1)+r+1} \equiv a^{(m_0-1)(q-2)+r+m_0}$$

$$\equiv a^{(m_0-1)(q-2)+r+1} \equiv \cdots \equiv a^{r+1} \pmod{b}.$$

When $q = 0$, we also have

$$a^{r+1} \equiv a \pmod{b}. \qquad \qquad ②$$

$r > 0$ would imply $m_0 > r+1 \geq 2$ and, by ②, contradicts the minimality of m_0. So $r = 0$ and $n \equiv 1 \pmod{m_0 - 1}$.

Similarly, we can prove $a^n \equiv a \pmod{b}$ when $n \equiv 1 \pmod{m_0 - 1}$. Hence ① is proved.

By ①, when $x \in S$ is the product of at least two elements in S, then x is also the product of m_0 elements in S. Indeed, when $x = x_1 x_2 \ldots x_l$, where $x_1, x_2, \ldots, x_l \in S$, $l > 1$, we have $a^l \equiv a \pmod{b}$, therefore $l \equiv 1 \pmod{m_0 - 1}$, i.e.,

$$l - m_0 + 1 \equiv 1 \pmod{m_0 - 1};$$

let $x' = x_{m_0} \ldots x_l$, then

$$x' \equiv a^{l-m_0+1} \equiv a \pmod{b},$$

so $x' \in S$, and x is indeed the product of m_0 elements in S, namely $x_1, \ldots, x_{m_0-1}, x'$.

Now suppose $(a, b) = p_1^{\alpha_1} p_2^{\alpha_2} \ldots p_k^{\alpha_k}$ is the standard factorization of (a, b), $k \geq 2$. We use the standard notation $\nu_p(n)$ to denote the highest power of a prime p in a positive integer n.

Note that, for $i = 1, 2, \ldots, k$, since $\nu_{p_i}((a, b)) = \alpha_i$ and $a^{m_0} \equiv a \pmod{b}$, $m_0 \geq 2$, it is easy to see that $\nu_{p_i}(b) = \alpha_i$.

Let $b = p_1^{\alpha_1} p_2^{\alpha_2} \ldots p_k^{\alpha_k} c$, where $(c, p_1 \ldots p_k) = 1$. For each $i = 1, 2, \ldots, k$, let $\delta_i = \delta_c(p_i)$ be the order of p_i modulo c. We prove that each element in S can be expressed as the product of up to

$$t = m_0 + \left\lfloor \frac{\delta_1 - 1}{\alpha_1} \right\rfloor + (m_0 - 1) \left\lfloor \frac{\delta_2 - 1}{\alpha_2} \right\rfloor$$

irreducible elements in S.

Note that, for a positive integer x, the condition $x \in S$ is equivalent to $x \equiv a \pmod{b}$, which is in turn equivalent to

$$\begin{cases} x \equiv a \equiv 0 \pmod{p_1^{\alpha_1} p_2^{\alpha_2} \ldots p_k^{\alpha_k}} \\ x \equiv a \pmod{c}, \end{cases}$$

which is also equivalent to

$$\begin{cases} \nu_{p_i}(x) \geq \alpha_i \ (1 \leq i \leq k), \\ x \equiv a \pmod{c}. \end{cases}$$

Combined with $p_i^{\delta_i} \equiv 1 \pmod{c}$ we have, when $x \in S$, $p_i^{\delta_i} x \in S$; when $x \in S$ and $\nu_{p_i}(x) \geq \alpha_i + \delta_i$, $p_i^{-\delta_i} x \in S$.

For any reducible element $x \in S$, we proved that x is the product of m_0 elements in S, let $x = y_1 y_2 \ldots y_{m_0}$, where $y_1, y_2, \ldots, y_{m_0} \in S$.

We may assume $\nu_{p_1}(y_1) \in [\alpha_1, \alpha_1 + \delta_1)$ — otherwise $\nu_{p_1}(y_1) \geq \alpha_1 + \delta_1$, we can pick positive integer u so that

$$\nu_{p_1}(y_1) - u\delta_1 \in [\alpha_1, \alpha_1 + \delta_1),$$

so $p_1^{-\delta_1 u} y_1, p_1^{\delta_1 u} y_2 \in S$ and we can replace y_1 and y_2 by $p_1^{-\delta_1 u} y_1$ and $p_1^{\delta_1 u} y_2$.

Similarly, for each i where $2 \leq i \leq m_0$, we may assume $\nu_{p_2}(y_i) \in [\alpha_2, \alpha_2 + \delta_2)$ — otherwise $\nu_{p_2}(y_i) \geq \alpha_2 + \delta_2$, we can pick positive integer u' so that

$$\nu_{p_2}(y_i) - u'\delta_2 \in [\alpha_2, \alpha_2 + \delta_2),$$

and we can replace y_i and y_1 by $p_2^{-\delta_2 u'} y_i$ and $p_2^{\delta_2 u} y_1$.

Since each element of S is divisible by $p_1^{\alpha_1}$, and $\nu_{p_1}(y_1) < \alpha_1 + \delta_1$, so y_1 is the product of at most $\left\lfloor \dfrac{\alpha_1 + \delta_1 - 1}{\alpha_1} \right\rfloor$ irreducible elements in S. Similarly, consider the power of p_2, each y_i, $2 \leq i \leq m_0$, can be expressed as the product of at most $\left\lfloor \dfrac{\alpha_2 + \delta_2 - 1}{\alpha_2} \right\rfloor$ irreducible elements from S. Therefore, $x = y_1 y_2 \ldots y_{m_0}$ can be expressed as the product of up to

$$\left\lfloor \frac{\alpha_1 + \delta_1 - 1}{\alpha_1} \right\rfloor + (m_0 - 1) \left\lfloor \frac{\alpha_2 + \delta_2 - 1}{\alpha_2} \right\rfloor = m_0 + \left\lfloor \frac{\delta_1 - 1}{\alpha_1} \right\rfloor + (m_0 - 1) \left\lfloor \frac{\delta_2 - 1}{\alpha_2} \right\rfloor$$

irreducible elements. □

Second Day

March 24, 2015

8:00–12:30

4 Given integer $n \geq 2$, and a sequence of monotone non-decreasing positive numbers $x_1, x_2, \ldots x_n$, such that $x_1, x_2/2, \ldots, x_n/n$ is monotone non-increasing. Prove that

$$\frac{A_n}{G_n} \leq \frac{n+1}{2 \sqrt[n]{n!}},$$

where A_n and G_n are the arithmetic and geometric mean of $x_1, x_2, \ldots, x_n$, respectively. (posed by Leng Gangsong)

Proof. From the statement we have $x_1 \leq x_2 \leq \cdots \leq x_n$, and

$$\frac{1}{x_1} \leq \frac{2}{x_2} \leq \cdots \leq \frac{n}{x_n}.$$

By Chebyshev's sum inequality,

$$\left(\frac{1}{n} \sum_{i=1}^{n} x_i \right) \left(\frac{1}{n} \sum_{i=1}^{n} \frac{i}{x_i} \right) \leq \frac{1}{n} \sum_{i=1}^{n} x_i \cdot \frac{i}{x_i} = \frac{n+1}{2}. \qquad \text{(1)}$$

By the AM-GM inequality,

$$\frac{1}{n} \sum_{i=1}^{n} \frac{i}{x_i} \geq \sqrt[n]{\frac{n!}{x_1 x_2 \ldots x_n}} = \frac{\sqrt[n]{n!}}{G_n}. \qquad \text{(2)}$$

By (1) and (2), we get

$$\frac{A_n}{G_n} \leq \frac{n+1}{2 \sqrt[n]{n!}}. \qquad \square$$

5 Colour each edge of the complete graph G with 2015 vertices with red and blue. Let V be the vertex set of G and define, for each 2-element subset $\{u, v\} \subseteq V$,

$$L(u, v) = \{u, v\} \cup \{w \in V : \text{there are exactly 2}$$

$$\text{red edges among } u, v, \text{ and } w\}.$$

Prove that, when $\{u, v\}$ is ranged over all 2-element subsets of V, there are at least 120 different $L(u, v)$'s. (posed by Yao Yijun)

Proof (First proof). Pick any vertex v. When we have 120 vertices in $V \setminus \{v\}$, say, $u_1, u_2, \ldots, u_{120}$, such that each vu_i is coloured blue, then, for any $i \neq j$, there are at least 2 blue edges among v, u_i, and u_j; it is easy to see $u_j \notin L(v, u_i)$. Also note that $u_i \in L(v, u_i)$, so u_i is the unique element among $u_1, \ldots, u_{120}$ that is contained in $L(v, u_i)$. This means $L(v, u_i)$, $1 \leq i \leq 120$, are 120 different sets, and we are done.

Otherwise, let W be the set of vertices u where vu is a red edge, we have

$$|W| \geq |V| - 1 - 119 = 2015 - 120 = 1895.$$

Consider all the $L(v, w)$, $w \in W$. We are done if there are at least 120 different such $L(v, w)$'s. Otherwise, by the pigeonhole principle, there are at least $\left\lceil \dfrac{|W|}{119} \right\rceil \geq \left\lceil \dfrac{1895}{119} \right\rceil = 16$ vertices in W, say $w_1, w_2, \ldots, w_{16}$, such that

$$L(v, w_1) = L(v, w_2) = \cdots = L(v, w_{16}).$$

For any $1 \leq i < j \leq 16$, $w_i \in L(v, w_i) = L(v, w_j)$, and note that vw_i and vw_j are both coloured red, so, by the definition of $L(v, w_j)$, $w_i w_j$ is coloured blue. So, for any 3 distinct elements $w, w', w'' \in W$, the three edges among them are all coloured blue, and $w'' \notin L(w, w')$. Consequently, for all $1 \leq i < j \leq 16$, $L(w_i, w_j)$ contains w_i, w_j, but none of the other elements from W. It follows that these $\dbinom{16}{2} = 120$ $L(w_i, w_j)$'s are different.

Proof (Based on the solution from Gao Jiyang in the test). We prove the stronger result that there are at least 1007 different $L(u, v)$'s.

Assume, for the sake of a contradiction, there are less than 1007 such sets.

If all the edges are red, all the $\dbinom{2015}{2} > 1007$ $L(u, v)$'s are different, contradicts our assumption. So blue edges exist, and let v_0 be the vertex that is incident to a maximum number (denoted by k) of blue edges, and assume $v_0 v_1, \ldots, v_0 v_k$ are coloured blue, and $v_0 v_{k+1}, \ldots, v_0 v_{2014}$ are coloured red.

Fact 1. $L(v_0, v_i) \neq L(v_0, v_j)$ for all $1 \leq i < j \leq k$.

Indeed, both $v_0 v_i$ and $v_0 v_j$ are blue edges, so $v_j \notin L(v_0, v_i)$, but $v_j \in L(v_0, v_j)$, hence $L(v_0, v_i) \neq L(v_0, v_j)$.

From Fact 1 and our assumption, $1 \leq k \leq 1006$.

Fact 2. $L(v_0, v_i) \neq L(v_0, v_j)$ for all $1 \leq i \leq k < j \leq 2014$.

Assume $L(v_0, v_i) = L(v_0, v_j)$ for some $1 \leq i \leq k < j \leq 2014$. Without loss of generality, $i = 1$ and $j = 2014$. For any l, where $2 \leq l \leq k$, $v_l \notin L(v_0, v_1) = L(v_0, v_{2014})$ implies $v_l v_{2014}$ is blue. By the maximality of k, v_{2014} is connected to at most one of $v_{k+1}, \ldots, v_{2013}$ by a blue edge. Without loss of generality, $v_{2014} v_{k+1}, \ldots, v_{2014} v_{2012}$ are all red. Then, for any $k+1 \leq l \leq 2012$, $v_l \notin L(v_0, v_{2014}) = L(v_0, v_1)$, which implies $v_1 v_l$ is blue. So v_1 is incident to at least

$$1 + (2012 - k) = 2013 - k \geq 2013 - 1006 = 1007 > k$$

blue edges, and contradicts the maximality of k.

Fact 3. If $L(v_0, v_i) = L(v_0, v_j)$ for some pair (i, j), where $k + 1 \leq i < j \leq 2014$, we must have vv_i and vv_j are coloured with the same colour for any $v \neq v_i, v_j$.

When $v = v_0$, vv_i and vv_j are both red and we are done. Otherwise $v \neq v_0$, when $v \in L(v_0 v_i) = L(v_0 v_j)$, there are exactly one red edge in vv_0 and vv_i, also exactly one red edge in vv_0 and vv_j, so vv_i and vv_j are of the same colour; when $v \notin L(v_0 v_i) = L(v_0 v_j)$, vv_0 and vv_i are both red or both blue, also vv_0 and vv_j are of the same colour, so vv_i and vv_j are of the same colour.

Fact 4. Suppose $L(v_0, v_i) = L(v_0, v_j) = L(v_0, v_l)$ for three distinct indices in $\{k + 1, \ldots, 2014\}$, then $L(u, v) \neq L(v_i, v_j)$ for any $\{u, v\} \neq \{v_i, v_j\}$.

Assume $L(u, v) = L(v_i, v_j)$. Since $v_i \in L(v_0, v_i) = L(v_0, v_j)$, $v_i v_j$ is blue. Similarly, $v_i v_l, v_j v_l$ are blue. So $v_l \notin L(v_i, v_j) = L(u, v)$. In particular, $v_l \neq u, v$. Since $\{u, v\} \neq \{v_i, v_j\}$, without loss of generality, $v_i \neq u, v$. By Fact 3, $v_i u$ and $v_l u$ are of the same colour, $v_i v$ and $v_l v$ are of the same colour, so $v_i \in L(u, v)$ if and only if $v_l \in L(u, v)$. However, $v_i \in L(v_i, v_j) = L(u, v)$ and $v_l \notin L(v_i, v_j) = L(u, v)$, a contradiction.

Having proved the facts above, now we consider the sequence of sets $L(v_0, v_i)$, $1 \leq i \leq 2014$. If the the same set appears three or more times, say, $L(v_0, v_i) = L(v_0, v_j) = L(v_0, v_l)$; by Facts 1 and 2, i, j, l are from $[k+1, 2014]$. Remove $L(v_0, v_i)$ and $L(v_0, v_j)$ from this sequence, and replace with two copies of $L(v_i, v_j)$. From Fact 4, these two sets are not the same as any other $L(u, v)$'s. Repeat the process until every set in this sequence appears at most twice, therefore we have at least 1007 different sets. $\square$

6 For any positive integer n, define $f(n) = \tau(n!) - \tau((n-1)!)$, where $\tau(a)$ is the number of positive divisors of a positive integer a. Prove that there are infinitely many composite numbers n, such that $f(m) < f(n)$ for any positive integer $m < n$. (posed by Yu Hongbing)

Proof (First proof). We prove the proposition in several steps.

Fact 1. $\tau((n-1)!) < \tau(n!)$ for any integer $n > 1$.

Since $(n-1)! \mid n!$, any positive divisor of $(n-1)!$ is also a divisor of $n!$. And $n!$ has at least one divisor, namely, $n!$, that does not divide $(n-1)!$. So $\tau((n-1)!) < \tau(n!)$.

Fact 2. $f(p) > f(m)$ for any odd prime p and integer $1 \le m < p$.

τ is a multiplicative function, so

$$\tau(p!) = \tau(p)\tau((p-1)!) = 2\tau((p-1)!).$$

Therefore,

$$f(p) = \tau(p!) - \tau((p-1)!) = 2\tau((p-1)!) - \tau((p-1)!) = \tau((p-1)!).$$

For $m < p$, by Fact 1, we have

$$f(m) = \tau(m!) - \tau((m-1)!) < \tau(m!) \le \tau((p-1)!) = f(p).$$

Fact 3. $f(2p) > f(q)$ for any odd primes p and q such that $p \le q < 2p$.

Suppose $(2p-1)! = 2^\alpha p A$, where $(A, 2p) = 1$, so

$$(2p)! = 2^{\alpha+1} p^2 A.$$

By the multiplicity of τ,

$$\frac{\tau((2p)!)}{\tau((2p-1)!)} = \frac{3(\alpha+2)}{2(\alpha+1)} > \frac{3}{2},$$

so

$$f(2p) = \tau((2p)!) - \tau((2p-1)!) > \frac{3}{2}\tau((2p-1)!) - \tau((2p-1)!) = \frac{1}{2}\tau((2p-1)!).$$

Yet $f(q) = \tau(q!) - \tau((q-1)!) = \frac{1}{2}\tau((q-1)!)$, together with Fact 1, we get $f(q) < f(2p)$.

Fact 4. For any odd prime p, there is an integer $n \in [p, 2p]$ satisfying the requirement in our problem. Since there are infinitely many primes, there are infinitely many such n's.

Let $n \in [p, 2p]$ be the least integer such that $f(n) = \max\{f(m) : p \leq m \leq 2p\}$. By Fact 3, n is composite. By this definition, $f(m) < f(n)$ for any $p \leq m < n$. In particular, $f(p) < f(n)$. For any $m < p$, we have, by Fact 2, $f(m) < f(p) < f(n)$.

Therefore $f(m) < f(n)$ for any $m < n$, and n satisfies the requirement.

Proof (Second proof). We prove that $n = 2p$ satisfies the requirement for any odd prime p.

We first prove the following lemma:

Lemma 5.1 $\tau(n!) \leq 2\tau((n-1)!)$ *for any positive integer* n.

Proof (of the lemma). The statement clearly holds for $n = 1$. Now assume $n > 1$ and

$$n = p_1^{\alpha_1} \cdots p_k^{\alpha_k}$$

is the standard factorization. For each $i = 1, 2, \ldots, k$, let β_i be the highest power of p_i in $n!$, then

$$\frac{\tau(n!)}{\tau((n-1)!)} = \prod_{i=1}^{k} \frac{\beta_i + 1}{\beta_i - \alpha_i + 1}. \qquad \textcircled{1}$$

Next we prove that, for each $i = 1, 2, \ldots, k$,

$$\frac{\beta_i + 1}{\beta_i - \alpha_i + 1} \leq 1 + \frac{p_i \alpha_i}{n}, \qquad \textcircled{2}$$

which is equivalent to $\dfrac{\alpha_i}{\beta_i - \alpha_i + 1} \leq \dfrac{p_i \alpha_i}{n}$, i.e.,

$$\beta_i \geq \frac{n}{p_i} + \alpha_i - 1.$$

Indeed, note that $p_i^{\alpha_i} \mid n$ and $p_i^{\alpha_i} \leq n$, we have

$$\beta_i = \sum_{j \geq 1} \left\lfloor \frac{n}{p_i^j} \right\rfloor \geq \sum_{j=1}^{\alpha_i} \frac{n}{p_i^j} = \frac{n}{p_i} + \sum_{j=2}^{\alpha_i} \frac{n}{p_i^j} \geq \frac{n}{p_i} + \sum_{j=2}^{\alpha_i} 1 = \frac{n}{p_i} + \alpha_i - 1.$$

Thus $\textcircled{2}$ is proved, and we get

$$\frac{\tau(n!)}{\tau((n-1)!)} = \prod_{i=1}^{k} \frac{\beta_i + 1}{\beta_i - \alpha_i + 1} \leq \prod_{i=1}^{k} \left(1 + \frac{p_i \alpha_i}{n}\right). \qquad \textcircled{3}$$

Let $n' = \prod_{i=1}^{k} p_i$. Since $\dfrac{p_i \alpha_i}{p_i^{\alpha_i}} \leq 1$, so $\dfrac{p_i \alpha_i}{n} \leq \dfrac{p_i}{n'}$, therefore

$$\prod_{i=1}^{k} \left(1 + \frac{p_i \alpha_i}{n}\right) \leq \prod_{i=1}^{k} \left(1 + \frac{p_i}{n'}\right). \qquad \text{④}$$

When $k = 1$, $1 + \dfrac{p_1}{n'} \leq 1 + 1 = 2$;

When $k = 2$,

$$\left(1 + \frac{p_1}{n'}\right)\left(1 + \frac{p_2}{n'}\right) \leq \left(1 + \frac{1}{p_2}\right)\left(1 + \frac{1}{p_1}\right) \leq \left(1 + \frac{1}{2}\right)\left(1 + \frac{1}{3}\right) = 2.$$

When $k \geq 3$, we have, for each $i = 1, 2, \ldots, k$,

$$\frac{n'}{p} = p_1 \ldots p_{i-1} p_{i+1} \ldots p_k \geq 2 \cdot 3^{k-2} \geq 2k,$$

so

$$\prod_{i=1}^{k} \left(1 + \frac{p_i}{n'}\right) \leq \left(1 + \frac{1}{2k}\right)^k = \sqrt{\left(1 + \frac{1}{2k}\right)^{2k}} < \sqrt{e} < 2.$$

So $\prod_{i=1}^{k} \left(1 + \frac{p_i}{n'}\right) \leq 2$, together with ③ and ④,

$$\frac{\tau(n!)}{\tau((n-1)!)} \leq \prod_{i=1}^{k} \left(1 + \frac{p_i}{n'}\right) \leq 2.$$

This completes the proof of the lemma.

For an odd prime p and $n = 2p$, the highest power of p in $n!$ and $(n-1)!$ are 2 and 1, respectively; also the highest power of 2 in $n!$ is bigger than the highest power of 2 in $(n-1)!$, so, by ①,

$$\frac{\tau(n!)}{\tau((n-1)!)} > \frac{3}{2},$$

i.e., $f(n) > \dfrac{1}{2}\tau((n-1)!)$.

Note that $\{\tau(n!)\}$ is strictly increasing, for any positive integer $m \leq n - 1$, by the lemma,

$$f(m) \leq \frac{1}{2}\tau(m!) \leq \frac{1}{2}\tau((n-1)!) < f(n).$$

So $n = 2p$ has the desired property.   □

China National Team Selection Test

First Day

March 25, 2016

8:00–12:30

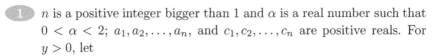 n is a positive integer bigger than 1 and α is a real number such that $0 < \alpha < 2$; $a_1, a_2, \ldots, a_n$, and $c_1, c_2, \ldots, c_n$ are positive reals. For $y > 0$, let

$$f(y) = \left(\sum_{a_i \leq y} c_i a_i^2 \right)^{\frac{1}{2}} + \left(\sum_{a_i > y} c_i a_i^{\alpha} \right)^{\frac{1}{\alpha}}.$$

Prove that, for any positive real x, $f(x) \leq 8^{1/\alpha} x$, if $x \geq f(y)$ for some y. (posed by Leng Gangsong)

Proof. We need a simple lemma.

Lemma 1 *For $X, Y \geq 0$ and $r > 0$,*

$$(X + Y)^r \leq 2^r (X^r + Y^r).$$

Proof.

$$(X + Y)^r \leq (2 \max\{X, Y\})^r = 2^r (\max\{X^r, Y^r\}) \leq 2^r (X^r + Y^r).$$

Thus completes the proof of the lemma.

73

Now suppose positive reals x and y satisfying $x \geq f(y)$.
When $y \geq x$, we have

$$f(x) = \left(\sum_{a_i \leq x} c_i a_i^2 \right)^{\frac{1}{2}} + \left(\sum_{a_i > x} c_i a_i^\alpha \right)^{\frac{1}{\alpha}}$$

$$\leq \left(\sum_{a_i \leq y} c_i a_i^2 \right)^{\frac{1}{2}} + \left(\sum_{x < a_i \leq y} c_i a_i^\alpha + \sum_{a_i > y} c_i a_i^\alpha \right)^{\frac{1}{\alpha}}. \qquad ①$$

By the lemma,

the right hand side of ①

$$\leq \left(\sum_{a_i \leq y} c_i a_i^2 \right)^{\frac{1}{2}} + 2^{\frac{1}{\alpha}} \left(\sum_{x < a_i \leq y} c_i a_i^\alpha \right)^{\frac{1}{\alpha}} + 2^{\frac{1}{\alpha}} \left(\sum_{a_i > y} c_i a_i^\alpha \right)^{\frac{1}{\alpha}} \qquad ②$$

$$\leq 2^{\frac{1}{\alpha}} f(y) + 2^{\frac{1}{\alpha}} \left(\sum_{x < a_i \leq y} c_i a_i^\alpha \right)^{\frac{1}{\alpha}}.$$

By the fact that $0 < \alpha < 2$ and $f(y) \leq x$, we have

$$\sum_{x < a_i \leq y} c_i a_i^\alpha = \sum_{x < a_i \leq y} c_i a_i^2 a_i^{\alpha - 2} \leq x^{\alpha - 2} \sum_{x < a_i \leq y} c_i a_i^2 \leq x^{\alpha - 2} f^2(y) \leq x^\alpha,$$

so, by ②,

$$f(x) \leq 2^{\frac{1}{\alpha}} f(y) + 2^{\frac{1}{\alpha}} x \leq 2^{\frac{1}{\alpha}} 2x \leq 8^{\frac{1}{\alpha}} x.$$

Similarly, when $y < x$, we have

$$f(x) = \left(\sum_{a_i \leq y} c_i a_i^2 + \sum_{y < a_i \leq x} c_i a_i^2 \right)^{\frac{1}{2}} + \left(\sum_{a_i > x} c_i a_i^\alpha \right)^{\frac{1}{\alpha}}$$

$$\leq 2^{\frac{1}{2}} \left(\sum_{a_i \leq y} c_i a_i^2 \right)^{\frac{1}{2}} + 2^{\frac{1}{2}} \left(\sum_{y < a_i \leq x} c_i a_i^2 \right)^{\frac{1}{2}} + \left(\sum_{a_i > y} c_i a_i^\alpha \right)^{\frac{1}{\alpha}}$$

$$\times \text{ (by the lemma)}$$

$$\leq 2^{\frac{1}{2}} f(y) + 2^{\frac{1}{2}} \left(\sum_{y < a_i \leq x} c_i a_i^2 \right)^{\frac{1}{2}}$$

$$\leq 2^{\frac{1}{2}} f(y) + 2^{\frac{1}{2}} \left(x^{2-\alpha} \sum_{y < a_i \leq x} c_i a_i^\alpha \right)^{\frac{1}{2}} \qquad \text{(since } 0 < \alpha < 2\text{)}$$

$$\leq 2^{\frac{1}{2}} f(y) + 2^{\frac{1}{2}} \left(x^{2-\alpha} f^\alpha(y) \right)^{\frac{1}{2}}$$

$$\leq 2^{\frac{1}{2}} x + 2^{\frac{1}{2}} x = 8^{\frac{1}{2}} x < 8^{\frac{1}{\alpha}} x.$$

In the last line above we used $0 < \alpha < 2$ and $f(y) \leq x$. $\qquad\square$

2 Call a point in the Cartesian coordinate system *rational* if both of its coordinates are rational. For any positive integer n, can all the rational points be coloured with n colours, one colour for each point, so that any segment with both ends rational contains all the n colours? (posed by Xiong Bin)

Proof. The answer is positive.

For a non-zero rational q, let $\nu(q)$ denote the power of 2 in q, to be specific, if $q = \dfrac{a}{b}$, where a and b are non-zero integers, $\nu(q) = \nu_2(a) - \nu_2(b)$. We also write $\nu(0) = \infty$.

Note that for any rationals x, y where $\nu(x) < \nu(y)$, we have

$$\nu(x + y) = \nu(x).$$

For any rational point $P = (x, y)$, where x and y are not both 0, let

$$\nu(P) = \min\{\nu(x), \nu(y)\},$$

and suppose $1 \leq i \leq n$ satisfies

$$\nu(P) \equiv i \pmod{n},$$

we colour P with the i-th colour. And colour the origin with an arbitrary colour.

We show that the above colouring satisfies our requirement.

Suppose A and B are two different rational points, $A = (x_0, y_0)$, then there are positive integers a and b, not both zero, such that all the rational points on the segment AB can be expressed as $(x_0 + at, y_0 + bt)$, where t ranges over all the rationals in $[0, \delta]$ for some positive rational δ.

For any positive integer m that is big enough, $2^{-m} \in [0, \delta]$. Without loss of generality, $\nu(a) \leq \nu(b)$, in particular, $a \neq 0$. For m big enough, $\nu(2^{-m}a) = \nu(a) - m < \min\{\nu(x_0), \nu(y_0)\}$. When this happens, $\nu(x_0 + 2^{-m}a) = \nu(a) - m$.

When $b \neq 0$, take m big enough so that $\nu(2^{-m}b) = \nu(b) - m < \nu(y_0)$, then $\nu(y_0 + 2^{-m}b) = \nu(b) - m$.

When $b = 0$, $\nu(y_0 + 2^{-m}b) = \nu(y_0)$.

In any case, there exists positive integer M big enough, such that for any $m \geq M$, $2^{-m} \in [0, \delta]$, and, for $P_m = (x_0 + 2^{-m}a, y_0 + 2^{-m}b)$, we have

$$\nu(P_m) = \min\{\nu(x_0 + 2^{-m}a), \nu(y_0 + 2^{-m}b)\} = \nu(a) - m.$$

Thus, when m ranges over all integers bigger than M, $\nu(P_m)$ takes all the residue classes modulo n, and AB contains all the n colours.   □

3 As in the figure, in an inscribed quadrilateral $ABCD$, $AB > BC$ and $AD > DC$; I and J are the incenters of $\triangle ABC$ and $\triangle ADC$, respectively. The circle with diameter AC intersects the segment IB at X and intersects the extension of JD at Y.

Prove that, when B, I, J, D are concyclic, the points X and Y are symmetric about AC. (posed by He Yijie)

Proof.   As in Figure 3.2, extend BI and DJ, let K be their intersection, and let M and N be their intersection with the circumcircle of the quadrilateral $ABCD$, respectively; connect M and N.

Clearly, M and N are the midpoints of the arcs $\overset{\frown}{ADC}$ and $\overset{\frown}{ABC}$, respectively, so MN is the diameter perpendicular to AC, the foot T is the midpoint of AC. Together with $AB > AC$ and $AD > DC$, we have B and D are on the same side of the line MN as C.

Since B, I, J, D are concyclic,

$$\angle MIJ = \angle JDB = \angle NDB = \angle NMB,$$

so $IJ \parallel MN$.

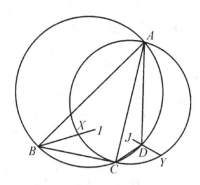

Fig. 3.1

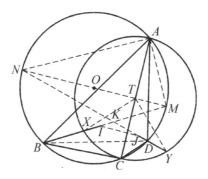

Fig. 3.2

By the property of the incenter,

$$NA = NJ, \quad MA = MI,$$

so

$$\frac{NT}{TY} = \frac{NT}{TA} = \cot \angle ANM = \frac{NA}{AM} = \frac{NJ}{MI} = \frac{NK}{KM}.$$

Therefore, by the law of sines,

$$\frac{\sin \angle TYN}{\sin \angle TNY} = \frac{NT}{TY} = \frac{NK}{KM} = \frac{\sin \angle KMN}{\sin \angle KNM}.$$

Note that $\angle TNY = \angle KNM$, so

$$\sin \angle TYN = \sin \angle KMN.$$

Clearly $\angle KMN = \angle BMN < 90°$, and by the fact that Y and N are on different sides of AC, we have

$$\angle TYN = \angle MTY - \angle TNY < 90°,$$

therefore $\angle TYN = \angle KMN$, and T, K, Y, and M are concyclic.

Similarly, T, K, X, and N are concyclic. So,

$$\angle MTY = \angle MKY = \angle NKX = \angle NTX.$$

Note that X and Y are on the different sides of AC, and the same side of MN, so the ray TX and TY are symmetric about AC; also X and Y are on the circle with diameter AC, so these two points are symmetric about AC. $\qquad\square$

Second Day

March 26, 2016

8:00–12:30

4 a, b, b', c, m, and q are positive integers, where $m > 1$ and $q > 1$, $|b - b'| \geq a$. There exists a positive integer M such that

$$S_q(an + b) \equiv S_q(an + b') + c \pmod{m} \qquad \text{①}$$

for all integers $n \geq M$, here $S_q(x)$ is the sum of the digits of a positive integer x in its base q representation.

Prove that ① holds for all positive integers n. (posed by Yu Hongbing)

Proof. Without loss of generality, $b > b'$, so $b - b' \geq a$, and there exists a positive integer r such that $b' < ar \leq b$.

Let $n = q^t - r$, where t is any integer big enough so that $q^t > ar - b'$ and $q^t > b - ar$. Then,

$$\begin{aligned}
&S_q(an + b) - S_q(an + b') \\
&= S_q(aq^t + b - ar) - S_q(aq^t - (ar - b')) \\
&= S_q(a) + S_q(b - ar) - S_q(aq^t - (ar - b')). \qquad \text{②}
\end{aligned}$$

On one hand, by the assumptions, for any integer $n = q^t - r \geq M$, the left hand side of ② is constant modulo m.

On the other hand, fix an integer s where $q^s > ar - b'$, for any $t > s+1$, and by the fact that

$$q^t > q^t - q^s > q^s > q^s - (ar - b'),$$

we get the right hand side of ② as

$$\begin{aligned}
&S_q(a) + S_q(b - ar) - S_q((a - 1)q^t + (q^t - q^s) + (q^s - (ar - b'))) \\
&= S_q(a) + S_q(b - ar) - S_q(a - 1) - S_q(q^t - q^s) - S_q(q^s - (ar - b')).
\end{aligned}$$

All the terms other than $S_q(q^t - q^s)$ on the right hand side of the above are constants.

So, for t big enough, $S_q(q^t - q^s)$ is constant modulo m. Note that the base q representation of $q^t - q^s$ is

$$q^t - q^s = (q - 1)q^{t-1} + (q - 1)q^{t-2} + \cdots + (q - 1)q^s,$$

so $S_q(q^t - q^s) = (q-1)(t-s)$ is a constant modulo m, therefore

$$m \mid (q-1).$$

Now, for any positive integer x, it is easy to see $S_q(x) \equiv x \pmod{m}$. So we have, for any positive integer n,

$$S_q(an+b) - S_q(an+b') \equiv (an+b) - (an+b') = b - b' \pmod{m}.$$

Since ① holds for all integers $n \geq M$, we must have $b - b' \equiv c \pmod{m}$, hence ① holds for all positive integers. □

5 S is a finite point set in the plane where any three points are not collinear, and its convex hull Ω is a 2016-gon $A_1 A_2 \ldots A_{2016}$. Label each point in S by one of the 4 numbers ± 1, ± 2, so that, for each $i = 1, 2, \ldots, 1008$, the labels on the points A_i and A_{i+1008} are opposite numbers.

Construct a set of triangles with vertices in S, so that any two of them do not share a common interior point, and the union of them is Ω. Prove that one of the triangles has two vertices labeled by opposite numbers. (posed by Yao Yijun)

Proof. The set of triangles in the statement form a triangulation $\mathcal{T}$ of the polygon Ω.

For the sake of a contradiction, we assume that none of the edges in $\mathcal{T}$ has opposite labels on its two ends.

For an edge e in $\mathcal{T}$ with two ends labeled by numbers in opposite signs, the labels must be $\{+1, -2\}$ or $\{-1, +2\}$, in the former case we call e a *type X* edge, in the latter case a *type Y* edge.

Let $e_1, e_2, \ldots, e_n$ be all the type X edges in $\mathcal{T}$.

We construct a simple graph G. The vertex set of G is $\{P_1, P_2, \ldots, P_n\}$, and for any $1 \leq i < j \leq n$, P_i and P_j are adjacent in G if and only if e_i and e_j are two edges of a triangle in $\mathcal{T}$.

Note that, when a triangle in $\mathcal{T}$ has a type X edge e, the third vertex of this triangle must be labeled with $+1$ or -2, so exactly one of the other two edges of the triangle is of type X.

So, when e_i lies on the edge of Ω, e_i is incident to exactly one triangle in $\mathcal{T}$, and the degree of P_i in G is 1; otherwise, e_i is incident to exactly two triangles in $\mathcal{T}$, and the degree of P_i in G is 2.

So G can be decomposed as vertex-disjoint cycles and paths, the end points, P_i and P_j, of each path correspond to two type X edges, e_i and e_j, on the boundary of Ω.

From here, all the indices are modulo 2016.

The number of type X edges in $A_i A_{i+1}$, $1 \leq i \leq 2016$, is even.

By the assumptions, A_i and A_{i+1008} are labeled with opposite numbers for each i; it follows from the definition of the type X and type Y edges that, for any i, $A_i A_{i+1}$ is of type Y if and only if $A_{i+1008} A_{i+1009}$ is of type X.

So the number of type Y edges among $A_i A_{i+1}$, where $1 \leq i \leq 1008$, equals the number of type X edges among $A_i A_{i+1}$, where $1009 \leq i \leq 2016$. So the number of type X edges and type Y edges in $A_i A_{i+1}$, $1 \leq i \leq 1008$, equals the total number of type X edges on the boundary of Ω, which is even.

So, start from A_1, go through $A_2, A_3, \ldots, A_{1008}$, and walk to A_{1009}, the signs changed an even number of times, which implies that A_1 and A_{1009} are labeled with the same sign, a contradiction.

So there must be a triangle where two vertices are labeled with opposite numbers. □

Remark. The problem setter noticed this problem during a talk in algebraic topology in Fudan University. The result and its generalization in higher dimensions is called Tucker's lemma, posed in Bulletin of AMS in 1944, and proved in the Proceedings of the First Canadian Mathematical Congress in 1945, and was used to prove the Lusternik-Schnirelmann and Borsuk-Ulam theorems in topology. The proof given here and its higher dimensional version was first published in 1981 by Freud and Todd in Journal of Combinatorial Theory, Series A, the title of the paper was A Constructive Proof of Tucker's Combinatorial Lemma.

⑥ Determine all functions $f : \mathbb{R}^+ \to \mathbb{R}^+$ satisfying: for any three distinct positive reals a, b, and c, three segments with lengths a, b, and c can form a triangle if and only if three segments with lengths $f(a)$, $f(b)$, and $f(c)$ can form a triangle. (posed by Qu Zhenhua)

Proof. We say u, v, and w form a triangle if the three segments with these lengths can form a triangle, otherwise we say u, v, and w do not form a triangle.

Suppose $f : \mathbb{R}^+ \to \mathbb{R}^+$ is a function satisfying our requirement. We prove the following.

(a) f is strictly increasing.

Assume there are positive reals a and b such that $a < b$ and $f(a) \geq f(b)$.

Pick a small enough positive ϵ so that $\frac{a}{2} + \epsilon < a$ and $\frac{a}{2} + \epsilon < \frac{b}{2}$. For any two distinct real numbers x and y in the interval $I = \left(\frac{a}{2}, \frac{a}{2} + \epsilon\right)$, x, y, and a form a triangle, and x, y, and b do not form a triangle. By our assumptions, $f(x), f(y), f(a)$ form a triangle, and $f(x), f(y)$, and $f(b)$ do not. So $f(x) + f(y) > f(a) \geq f(b)$, therefore, at most one of $f(x) \leq f(b)$ and $f(y) \leq f(b)$ holds. It follows that there is at most one $x \in I$ where $f(x) \leq f(b)$.

Let $I_1 = \{x \in I : f(x) > f(b)\}$, from the discussion above, I_1 is an infinite set. Consider $f(I_1) = \{f(x) : x \in I_1\}$, we discuss two cases.

Case 1. $f(I_1)$ is bounded from above. Let M be an upper bound. For any $x \neq y \in I_1$, since $f(x)$, $f(y)$, and $f(b)$ do not form a triangle,

$$|f(x) - f(y)| \geq f(b) > 0,$$

so there are at most $\left\lfloor \dfrac{M}{f(b)} \right\rfloor$ elements in $f(I_1)$, and each of them has a unique pre-image in I_1, therefore I_1 is finite, contradicts the fact that I_1 is infinite.

Case 2. $f(I_1)$ is not bounded from above. Then there exists $x \in I_1$ such that $f(x) > f(a)$, and in turn there exists $y \in I_1$ such that $f(y) > f(x) + f(a)$, contradicts with the fact that $f(x)$, $f(y)$, and $f(a)$ form a triangle.

So we have proved that f is strictly monotone increasing.

(b) For any strictly decreasing series of positive reals $\{x_n\}$ such that $\lim_{n \to \infty} x_n = 0$, we have $\lim_{n \to \infty} f(x_n) = 0$.

From (a) we know that $\{f(x_n)\}$ is also strictly decreasing, so $\lim_{n \to \infty} f(x_n)$ exists. Let $A = \lim_{n \to \infty} f(x_n) \geq 0$.

Now assume $A > 0$, then there is a positive integer N so that $A < f(x_n) < 2A$ for all $n \geq N$. By the assumption on $\{x_n\}$, there are indices i, j, and k such that $N \leq i < j < k$ and $x_k < x_j < \frac{x_i}{2}$. Now x_i, x_j, and x_k are distinct positive reals that do not form a triangle; yet $f(x_i), f(x_j)$, and $f(x_k)$ do form a triangle because they are all in the interval $(A, 2A)$, a contradiction.

So $\lim_{n \to \infty} f(x_n) = 0$.

(c) For any strictly decreasing series of positive reals $\{x_n\}$ such that $\lim_{n \to \infty} x_n = y > 0$, we have $\lim_{n \to \infty} f(x_n) = f(y)$.

By (a), $\{f(x_n)\}$ is also strictly decreasing and $f(x_n) > f(y)$, so $\lim_{n\to\infty} f(x_n)$ exists. Let $A = \lim_{n\to\infty} f(x_n) \geq y$.

Consider $x_{n+1}, y, x_n - y$, by the assumptions on $\{x_n\}$, when n is big enough, $0 < x_n - y < y < x_{n+1}$; yet $(x_n - y) + y = x_n > x_{n+1}$, so they form a triangle. Therefore,

$$f(x_n - y) + f(y) > f(x_{n+1}). \qquad \text{⑤}$$

Since $\lim_{n\to\infty}(x_n - y) = 0$, by (b), $\lim_{n\to\infty} f(x_n - y) = 0$. Taking the limit with $n \to \infty$ on both sides of ⑤ , we have $f(y) \geq \lim_{n\to\infty} f(x_{n+1}) = A$. So $\lim_{n\to\infty} f(x_n) = A = f(y)$.

(d) $f(x + y) = f(x) + f(y)$ for any $x, y > 0$.

Take a strictly decreasing series $\{y_n\}$ such that $\lim_{n\to\infty} y_n = y$ and each term is less than $x + y$ but does not equal x. So x, y_n, and $x + y$ are distinct positive reals that form a triangle, therefore $f(x) + f(y_n) > f(x+y)$. Taking the limit with $n \to \infty$ and by (c) we get $f(x) + f(y) \geq f(x + y)$.

On the other hand, x, y_n, and $x + y_n$ are three distinct reals that do not form a triangle, so $f(x) + f(y_n) \leq f(x + y_n)$. Taking the limit when $n \to \infty$, we get $f(x) + f(y) \leq f(x + y)$.

So $f(x) + f(y) = f(x + y)$.

(e) When $f(1) = c > 0$, $f(x) = cx$ for all x.

By (d), it is easy to deduce that $f(q) = qf(1) = cq$ for any positive rational q.

For any positive real x, take a series of strictly decreasing rationals $\{q_n\}$ such that $\lim_{n\to\infty} q_n = x$. By (c),

$$f(x) = \lim_{n\to\infty} f(q_n) = \lim_{n\to\infty} cq_n = cx.$$

It is clear that, for any $c > 0$, $f(x) = cx$ satisfies our requirement. So these are all the functions we seek. $\qquad \square$

China Girls' Mathematical Olympiad

2014 (Zhongshan, Guangdong)

First Day

August 12, 2014

8:00–12:00

1. As shown in Figure 1.1, circles $\odot O_1$ and $\odot O_2$ intersect at points A and B, ray O_1A meets $\odot O_2$ at point C, and ray O_2A meets $\odot O_1$ at point D. Line BE through B and parallel to O_2A meets $\odot O_1$ at another point E. If $DE \parallel O_1A$, prove that $DC \perp CO_2$. (posed by Zheng Huan)

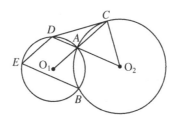

Fig. 1.1

Proof. As in Figure 1.2, since $\angle O_1DA = \angle O_1AD = \angle O_2AC = \angle O_2CA$, so the points O_1, O_2, C, D are concyclic, and

$$\angle DO_1O_2 + \angle DCO_2 = 180°.$$

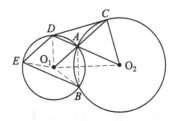

Fig. 1.2

Since $BE \parallel AD$, so $AB = DE$, therefore $\angle EDO_1 = \angle O_1AB$.

And since $DE \parallel O_1A$, so $\angle DO_1A = \angle EDO_1 = \angle O_1AB$, therefore $DO_1 \parallel AB$.

Since $AB \perp O_1O_2$, so $DO_1 \perp O_1O_2$, and $\angle DO_1O_2 = 90°$.

Hence, $\angle DCO_2 = 90°$, $DC \perp CO_2$.   □

2 For a given integer $n \geq 2$, n reals $x_1, x_2, \ldots, x_n$ satisfy the condition that $\lfloor x_1 \rfloor, \lfloor x_2 \rfloor, \ldots, \lfloor x_n \rfloor$ is a permutation of $1, 2, \ldots, n$. Determine the maximum and minimum values of $\sum_{i=1}^{n-1} \lfloor x_{i+1} - x_i \rfloor$. (posed by Liang Yingde)

Solution $\lfloor x_1 \rfloor, \lfloor x_2 \rfloor, \ldots, \lfloor x_n \rfloor$ is a permutation of $1, 2, \ldots, n$, so $1 \leq x_i < n + 1$ for any $1 \leq i \leq n$. In particular, $1 \leq x_1 < n + 1$ and $1 < x_n < n + 1$, hence

$$-n < x_n - x_1 < n.$$

Let $S = \sum_{i=1}^{n-1} \lfloor x_{i+1} - x_i \rfloor$. Since

$$\lfloor x_{i+1} - x_i \rfloor \leq x_{i+1} - x_i,$$

we have

$$S \leq \sum_{i=1}^{n-1} (x_{i+1} - x_i) = x_n - x_1 < n.$$

Note that S is an integer, so $S \leq n - 1$.

On the other hand, $\lfloor x_{i+1} - x_i \rfloor > x_{i+1} - x_i - 1$, so

$$S > \sum_{i=1}^{n-1} (x_{i+1} - x_i - 1) = x_n - x_1 - n + 1 > 1 - 2n.$$

Again, being an integer, $S \geq 2 - 2n$.

Take $x_i = i$ $(i = 1, 2, \ldots, n)$, we get $S = n - 1$.

Take $x_i = n + 1 - i + \dfrac{1}{i+1}$ $(i = 1, 2, \ldots, n)$, for any $i = 1, 2, \ldots, n-1$, we have

$$\lfloor x_{i+1} - x_i \rfloor = \left\lfloor \left(n + 1 - (i+1) + \frac{1}{i+2} \right) - \left(n + 1 - i + \frac{1}{i+1} \right) \right\rfloor$$

$$= \left\lfloor -1 - \frac{1}{(i+1)(i+2)} \right\rfloor = -2.$$

So $S = 2 - 2n$.

In conclusion, the maximum and minimum values for $\sum_{i=1}^{n-1} \lfloor x_{i+1} - x_i \rfloor$ are $n - 1$ and $2 - 2n$, respectively. □

3 In a group of n students, any two persons either know each other or do not know each other, and each student knows exactly d female students and d male students. Determine all such possible pair (n, d) of positive integers. (posed by Wang Xinmao)

Solution Construct a graph $G = (V, E)$ where V is the set of the students and two students are connected by an edge if they know each other. Suppose there are x girls and y boys. It is clear that $x, y > d$. The number of edges between the boys and the girls can be counted in two ways as xd and yd, so $x = y = n/2 \geq d + 1$. The number of edges between the girls is $\dfrac{xd}{2} = \dfrac{nd}{4}$. Therefore, n and d must satisfy

$$\frac{n}{2} \in \mathbb{N}^*, \quad \frac{nd}{4} \in \mathbb{N}^*, \quad n \geq 2d + 2.$$

We show that, for every such pair of positive integers (n, d), there is a graph that meets our condition.

Let $n = 2m$, the set of the girls $A = \{u_1, u_2, \ldots, u_m\}$, and the set of boys $B = \{v_1, v_2, \ldots, v_m\}$. The vertex set $V = A \cup B$.

When d is odd,

$$E = \{u_i u_j, v_i v_j, u_i v_j : i - j \in \{\pm 1, \pm 2, \ldots, \pm (d-1)/2, m/2\} \bmod m\};$$

and when d is even,

$$E = \{u_i u_j, v_i v_j, u_i v_j : i - j \in \{\pm 1, \pm 2, \ldots, \pm d/2\} \bmod m\}.$$ □

4 For every integer $m \geq 4$, define T_m to be the number of integer sequences $a_1, a_2, \ldots, a_m$ satisfying the following conditions:

(1) $a_i \in \{1, 2, 3, 4\}$, for all $i = 1, 2, \ldots, m$;
(2) $a_1 = a_m = 1$, $a_2 \neq 1$;
(3) $a_i \neq a_{i-1}$ and $a_i \neq a_{i-2}$, for all $i = 3, 4, \ldots, m$.

Prove that there exists a geometric sequence $\{g_n\}$ with positive terms such that, for any integer $n \geq 4$, $g_n - 2\sqrt{g_n} < T_n < g_n + 2\sqrt{g_n}$. (posed by Zhu Huawei)

Proof (First proof). Enumerate for small m's, we get $T_4 = T_5 = T_6 = 6$. For $n \geq 7$, we count such sequences $(a_1, a_2, \ldots, a_n)$ by dividing them into two cases.

Case 1. None of $a_2, a_3, \ldots, a_{n-1}$ equals 1. There are 3 choices for a_2, then 2 choices for a_3, and then for each $a_4, a_5, \ldots, a_{n-1}$, in this order, we must pick the only one among $\{2, 3, 4\}$ that differs from the two elements that precede it; finally we pick $a_n = 1$. Hence, there are exactly 6 sequences in this case.

Case 2. Some of $a_2, a_3, \ldots, a_{n-1}$ equals 1. Let k be the smallest index in $\{2, 3, \ldots, n-1\}$ such that $a_k = 1$. By (3), $4 \leq k \leq n-3$. Similar to Case 1, there are 6 different choices for $(a_2, a_3, \ldots, a_{k-1})$. Consider $b_1 = a_k$, $b_2 = a_{k+1}$, $\ldots$, $b_{n-k+1} = a_n$. In addition to the three conditions in the statement, the sequence $(b_1, b_2, \ldots, b_{n-k+1})$ satisfies $b_2 \neq a_{k-1}$. Notice that the three numbers 2, 3, and 4 are symmetric in our problem, we conclude that there are $\dfrac{2}{3} T_{n-k+1}$ choices for $(b_1, b_2, \ldots, b_{n-k+1})$. So, we have

$$\sum_{k=4}^{n-3} 6 \cdot \frac{2}{3} T_{n-k+1} = 4 \sum_{k=4}^{n-3} T_{n-k+1} = 4 \sum_{l=4}^{n-3} T_l$$

different sequences for Case 2.

Thus, when $n \geq 7$, we have

$$T_n = 6 + 4 \sum_{l=4}^{n-3} T_l.$$

Substitute n with $n+1$,

$$T_{n+1} = 6 + 4 \sum_{l=4}^{(n+1)-3} T_l = T_n + 4 T_{n-2}.$$

Note that $T_7 = 6 + 4T_4 = T_6 + 4T_4$, so $T_n = T_{n-1} + 4T_{n-3}$ holds for any $n \geq 7$. The characteristic equation for the linear recurrence is $x^3 = x^2 + 4$; its roots are $x_1 = 2$, $x_2 = \dfrac{-1 + \sqrt{7}i}{2}$, $x_3 = \dfrac{-1 - \sqrt{7}i}{2}$. The solution for T_n is in the form $T_n = Ax_1^n + Bx_2^n + Cx_3^n$, where A, B, and C are constants to be determined. With $T_4 = T_5 = T_6 = 6$, we get

$$A = \frac{3}{16}, \quad B = \frac{6\sqrt{7}i}{7}x_2^{-5}, \quad C = -\frac{6\sqrt{7}i}{7}x_3^{-5}.$$

So,

$$T_n = 3 \cdot 2^{n-4} + \frac{6\sqrt{7}i}{7}\left(x_2^{n-5} - x_3^{n-5}\right).$$

Let $\{g_n\}$ be the geometric sequence where $g_n = 3 \cdot 2^{n-4}$. For any $n \geq 3$, we have

$$|T_n - g_n| = \left|\frac{6\sqrt{7}i}{7}\left(x_2^{n-5} - x_3^{n-5}\right)\right| \leq \frac{6\sqrt{7}}{7}\left(\left|x_2^{n-5}\right| + \left|x_3^{n-5}\right|\right)$$

$$\leq \frac{6\sqrt{7}}{7} \cdot 2 \cdot \sqrt{2}^{n-5} = 2\sqrt{\frac{9}{7} \cdot 2^{n-3}} = 2\sqrt{\frac{6}{7}g_n} < 2\sqrt{g_n}.$$

Therefore, $g_n - 2\sqrt{g_n} < T_n < g_n + 2\sqrt{g_n}$. □

Proof (Second proof). It is easy to check that $T_4 = T_5 = 6$. For $n \geq 6$, count the sequences $(a_1, a_2, \ldots, a_{n-1})$ satisfying

(A1) $a_i \in \{1, 2, 3, 4\}$, for all $i = 1, 2, \ldots, n-1$;
(A2) $a_1 = 1$, $a_2 \neq 1$; and
(A3) $a_i \neq a_{i-1}$ and $a_i \neq a_{i-2}$, for all $i = 3, 4, \ldots, n-1$.

We successively chose $a_2, a_3, \ldots, a_{n-1}$. We have 3 choices for a_2, then for each $a_3, a_4, \ldots, a_{n-1}$ we have 2 choices. The number of such sequences is $3 \cdot 2^{n-3}$. On the other hand, these sequences can be partitioned into 3 types: (i) $a_{n-2} = 1$; (ii) $a_{n-1} = 1$; and (iii) $a_{n-2} \neq 1$, $a_{n-1} \neq 1$. Each of the T_{n-2} sequence satisfying the problem statement for $m = n - 2$ corresponds to two type (i) sequences — by adding a_{n-1} that is different from a_{n-2} and a_{n-3}. So, there are $2T_{n-2}$ sequences in type (i). Type (ii) sequences are exactly those T_{n-1} satisfying the problem statement for $m = n - 1$. By adding $a_n = 1$ to each sequence in type (iii), we get those T_n sequences satisfying the problem statement for $m = n$. Thus,

$$T_n + T_{n-1} + 2T_{n-2} = 3 \cdot 2^{n-3}.$$

Let $U_n = T_n - 3 \cdot 2^{n-4}$, we have $U_n + U_{n-1} + 2U_{n-2} = 0$, and $U_4 = 3$, $U_5 = 0$. The characteristic equation for the linear recurrence of U_n is $x^2 + x + 2 = 0$, has roots $\dfrac{-1 \pm \sqrt{7}i}{2}$. With $U_4 = 3$ and $U_5 = 0$, we get

$$U_n = \frac{6\sqrt{7}i}{7}\left(\left(\frac{-1+\sqrt{7}i}{2}\right)^{n-5} - \left(\frac{-1-\sqrt{7}i}{2}\right)^{n-5}\right).$$

The rest of the proof is the same as the last part of the first proof.　　□

Second Day

August 13, 2014

8:00–12:00

5 Let a be a non-square positive integer, and let r be a real root of the equation $x^3 - 2ax + 1 = 0$. Prove that $r + \sqrt{a}$ is irrational. (posed by Li Shenghong)

Proof. Assume $q = r + \sqrt{a}$ is rational. Substitute $r = q - \sqrt{a}$ in the equation, we get

$$(q - \sqrt{a})^3 - 2a(q - \sqrt{a}) + 1 = 0.$$

i.e.,

$$q^3 + aq + 1 - (3q^2 - a)\sqrt{a} = 0.$$

Since a is not a square, $\sqrt{a}$ is not rational, so

$$\begin{cases} q^3 + aq + 1 = 0, \\ 3q^2 - a = 0. \end{cases}$$

Canceling a, we get $4q^3 + 1 = 0$, which contradicts the assumption that q is rational.

6 As shown in Figure 6.1, in an acute $\triangle ABC$, $AB > AC$, D and E are the midpoints of sides AB and AC, respectively. The circumcircles of $\triangle ADE$ and $\triangle BCE$ meet at distinct points P and E; the circumcircles of $\triangle ADE$ and $\triangle BCD$ meet at distinct points Q and D. Prove that $AP = AQ$. (posed by Fu Yunhao)

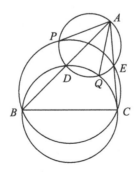

Fig. 6.1

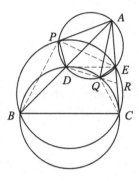

Fig. 6.2

Proof (First proof). As in Figure 6.2, connect DE, PD, QE, PB, QC, PE, and QD. Let R be the intersection of the lines QD and AC.

By the inscribed angle theorems, we have

$$\angle APD = \pi - \angle AED = \pi - \angle ACB;$$

$$\angle BPD = \angle BPE - \angle EPD = (\pi - \angle ACB) - \angle BAC = \angle ABC;$$

$$\angle AQE = \angle ADE = \angle ABC;$$

$$\angle CQE = \angle CQR + \angle RQE = \angle ABC + \angle DAE = \pi - \angle ACB.$$

So,

$$\angle APB = \angle APD + \angle BPD = \angle AQE + \angle CQE = \angle AQC.$$

One the other hand, in the triangle APB,

$$\frac{AP}{BP} = \frac{AP}{AD} \cdot \frac{BD}{BP} = \frac{\sin \angle ADP}{\sin \angle APD} \cdot \frac{\sin \angle BPD}{\sin \angle BDP} = \frac{\sin \angle BPD}{\sin \angle APD}$$

$$= \frac{\sin \angle ABC}{\sin(\pi - \angle ACB)}.$$

Similarly, we have $\dfrac{CQ}{AQ} = \dfrac{\sin \angle ABC}{\sin(\pi - \angle ACB)}$. So, $\dfrac{AP}{BP} = \dfrac{CQ}{AQ}$. Combined with the fact $\angle APB = \angle AQC$, this implies $\triangle APB \sim \triangle CQA$. Because D and E are the midpoints of the corresponding edges in these two triangles, so, $\triangle APD \sim \triangle CQE$. Therefore $\angle ADP = \angle CEQ = \angle ADQ$, and $AP = AQ$.

Proof (Second proof). As in Figure 6.3, consider an inversion with respect to a circle with center A and any radius, denote the images of B, C, D, E, P, Q under the inversion as B', C', D', E', P', Q', respectively.

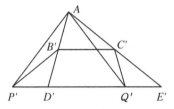

Fig. 6.3

By the properties of the inversions, $B'C'$ is the midline of $\triangle AD'E'$; points P' and Q' are on the line $D'E'$, the point P' is on the circumcircle of $\triangle B'C'E'$, and Q' is on the circumcircle of $\triangle B'C'D'$.

Because $B'C' \parallel D'E'$, both $B'C'Q'D'$ and $B'C'E'P'$ are isosceles trapezoids. So, $C'Q' = B'D' = B'A$, $C'A = C'E' = B'P'$. And we have

$$\angle AB'P' = \pi - \angle D'B'P' = \pi - (\angle P'B'C' - \angle D'B'C')$$

$$= \pi - (\angle E'C'B' - \angle Q'C'B') = \pi - \angle Q'C'E' = \angle Q'C'A.$$

So, $\triangle AB'P' \cong \triangle Q'C'A$, and $AP' = AQ'$. Therefore $AP = AQ$. □

Proof (Third proof). $\triangle ADE$ and $\triangle ABC$ are homothetic with external center A, their circumcircles are externally tangent with A as the point of tangency. Their common tangent line through A meets BC at a point M. AM is the radical axis of these two circles, BC is the radical axis of the circumcircles of $\triangle ABC$ and $\triangle DBC$, so, M is the power center of the three circles, thus is on the line PE. Similarly, M is on the line DQ. (Figure 6.4)

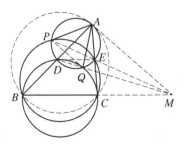

Fig. 6.4

By the property of the tangent line, $\angle CAM = \angle ABM$, so $\triangle ACM \sim \triangle BAM$, hence

$$\frac{BM}{AM} = \frac{AB}{AC} = \frac{BD}{AE}.$$

Also, $\angle DBM = \angle EAM$, so $\triangle BDM \sim \triangle AEM$ and we have $\angle AME = \angle BMD$.

Since $DE \parallel BC$, $\angle BMD = \angle MDE = \angle QPE$; so, $\angle AME = \angle QPE$, which implies $PQ \parallel AM$.

Therefore $\angle PQA = \angle QAM = \angle APQ$, and $AP = AQ$. $\square$

7 For any non-empty finite set X of real numbers, denote by $|X|$ the size of X, and define $f(X) = \frac{1}{|X|} \sum_{a \in X} a$. Let (A, B) be a pair of sets such that $A \cup B = \{1, 2, \ldots, 100\}$, $A \cap B = \emptyset$, and $1 \le |A| \le 98$. For any $p \in B$, let $A_p = A \cup \{p\}$ and $B_p = \{x \in B : x \ne p\}$. Find the maximum value of

$$(f(A_p) - f(A))(f(B_p) - f(B))$$

over all such (A, B) and all $p \in B$. (posed by He Yijie)

Solution   Let $S = (f(A_p) - f(A))(f(B_p) - f(B))$. By the definition of f, A_p, and B_p, we have

$$(|A| + 1)f(A_p) = |A|f(A) + p,$$

$$(|B| - 1)f(B_p) = |B|f(B) - p.$$

So,

$$S = \left(\frac{|A|f(A) + p}{|A| + 1} - f(A) \right) \left(\frac{|B|f(B) - p}{|B| - 1} - f(B) \right)$$

$$= \frac{(p - f(A))(f(B) - p)}{(|A| + 1)(|B| - 1)} \tag{1}$$

and

$$S = \left(f(A_p) - \frac{(|A| + 1)f(A_p) - p}{|A|} \right) \left(f(B_p) - \frac{(|B| - 1)f(B_p) + p}{|B|} \right)$$

$$= \frac{(p - f(A_p))(f(B_p) - p)}{|A| \cdot |B|}. \tag{2}$$

We prove that, $|f(A) - f(B)| \le 50$, $|f(A_p) - f(B_p)| \le 50$, and $S \le \frac{625}{196}$.

Notice that $|A| + |B| = 100$, we have

$$f(A) - f(B) \leq \frac{100 + 99 + \cdots + (|B| + 1)}{|A|} - \frac{1 + 2 + \cdots + |B|}{|B|}$$

$$= \frac{100 + |B| + 1}{2} - \frac{|B| + 1}{2} = 50.$$

$$f(A) - f(B) \geq \frac{1 + 2 + \cdots + |A|}{|A|} - \frac{100 + 99 + \cdots + (|A| + 1)}{|B|}$$

$$= \frac{1 + |A|}{2} - \frac{100 + (|A| + 1)}{2} = -50.$$

Therefore, $|f(A) - f(B)| \leq 50$. By the same reason, because $|A_p| + |B_p| = 100$, $|f(A_p) - f(B_p)| \leq 50$.

When $1 \leq |A| \leq 97$, since

$$(p - f(A))(f(B) - p) \leq \left(\frac{f(B) - f(A)}{2} \right)^2 \leq 625,$$

and $(|A| + 1)(|B| - 1) \geq 2 \times 98 = 196$, by $\boxed{1}$ we get

$$S = \frac{(p - f(A))(f(B) - p)}{(|A| + 1)(|B| - 1)} \leq \frac{625}{196}.$$

When $|A| = 98$, by $\boxed{2}$ we get

$$S = \frac{(p - f(A_p))(f(B_p) - p)}{|A| \cdot |B|} \leq \frac{1}{196} \left(\frac{f(B) - f(A)}{2} \right)^2 \leq \frac{625}{196}.$$

On the other hand, take $A = \{1\}$, $B = \{2, 3, \ldots, 100\}$, and $p = 26$. By $\boxed{1}$ we can check that

$$S = \frac{(26 - 1)(51 - 26)}{2 \times 98} = \frac{625}{196}.$$

Thus the maximum value for S is $\dfrac{625}{196}$. $\qquad\square$

8 Let n be a positive integer, and S be the set of integers in $\{1, 2, \ldots, n\}$ that are coprime to n. Let $S_1 = S \cap \left(0, \dfrac{n}{3}\right]$, $S_2 = S \cap \left(\dfrac{n}{3}, \dfrac{2n}{3}\right]$, and $S_3 = S \cap \left(\dfrac{2n}{3}, n\right]$. Prove that, when the number of elements in S is divisible by 3, S_1, S_2, and S_3 have the same number of elements. (posed by Wang Bin)

Proof. Denote $|X|$ the number of elements in a finite set X. For any positive integer n, define $A(n)$ to be the set of positive integers that are coprime to n; and for each positive integer k, define $A_k(n) = A(n) \cap \left(\dfrac{k-1}{3}n, \dfrac{k}{3}n \right]$.
For any x, $(x, n) = 1 \Leftrightarrow (x+n, n) = 1$, so $x \in A_k(n) \Leftrightarrow x+n \in A_{k+3}(n)$, and $|A_k(n)| = |A_{k+3}(n)|$.

If n satisfies the conclusion in our problem, $|A_1(n)| = |A_2(n)| = |A_3(n)|$, it is equivalent to the assertion that $A_k(n)$ are the same for all k, and we call such n *balanced*.

Now we prove the

Lemma 7.1 *If n is a balanced number and p is any prime, then pn is balanced.*

Proof. By the definition, $A_k(n)$ are all equal to the same number, denoted by m. We consider two situations.

Case 1. $p \mid n$. Then $(x, pn) = 1 \Leftrightarrow (x, n) = 1$. $A(pn) = A(n)$.

$$
\begin{aligned}
A_1(pn) &= A(pn) \cap \left(0, \frac{1}{3}pn \right] = A(n) \cap \left(0, \frac{p}{3}n \right] \\
&= A_1(n) \cup A_2(n) \cup \cdots \cup A_p(n), \\
A_2(pn) &= A(pn) \cap \left(\frac{1}{3}pn, \frac{2}{3}pn \right] = A(n) \cap \left(\frac{p}{3}n, \frac{2p}{3}n \right] \\
&= A_{p+1}(n) \cup A_{p+2}(n) \cup \cdots \cup A_{2p}(n), \\
A_3(pn) &= A(pn) \cap \left(\frac{2}{3}pn, pn \right] = A(n) \cap \left(\frac{2p}{3}n, \frac{3p}{3}n \right] \\
&= A_{2p+1}(n) \cup A_{2p+2}(n) \cup \cdots \cup A_{3p}(n),
\end{aligned}
$$

Thus $|A_1(pn)| = |A_2(pn)| = |A_3(pn)| = pm$, and pn is balanced.

Case 2. $p \nmid n$. $(x, pn) = 1 \Leftrightarrow (x, n) = 1 \wedge (x, p) = 1$. So, $A(pn) = A(n) \setminus B(n)$, where

$$
B(n) = \{ x \in A(n) : p \mid x \} = \{ x : x = py, y \in A(n) \}.
$$

By the discussion in Case 1, $A(n)$ has pm elements in each of the intervals $\left(0, \frac{1}{3}pn \right]$, $\left(\frac{1}{3}pn, \frac{2}{3}pn \right]$, and $\left(\frac{2}{3}pn, pn \right]$. Each element $x = py \in \left(0, \frac{1}{3}pn \right]$ of $B(n)$ corresponds to an element $y \in \left(0, \frac{1}{3}n \right]$ of $A(n)$. There are

$|A_1(n)| = m$ such y's, hence m elements of $B(n)$ in the interval $\left(0, \frac{1}{3}pn\right]$. Similarly, $B(n)$ has m elements in the other two intervals. So,

$$|A_1(pn)| = |A_2(pn)| = |A_3(pn)| = pm - m = (p-1)m,$$

and pn is balanced.

This completes the proof of the lemma.

Repeatedly using the lemma, we conclude that any multiple of a balanced number is itself balanced.

Now, suppose $|S|$ is a multiple of 3, factorize n as $p_1^{\alpha_1} p_2^{\alpha_2} \ldots p_k^{\alpha_k}$, where $p_1, p_2, \ldots, p_k$ are different primes, and $\alpha_1, \alpha_2, \ldots, \alpha_k$ are positive integers. Since

$$3 \mid |S| = \Phi(n) = (p_1 - 1)(p_2 - 1) \ldots (p_k - 1) p_1^{\alpha_1 - 1} p_2^{\alpha_2 - 1} \ldots p_k^{\alpha_k - 1},$$

either n is a multiple of 9 or else it has a prime factor in the form $3s + 1$.

It is easy to verify that 9 is balanced, as well as any prime in the form $3s + 1$, so n is balanced. $\qquad \square$

China Girls' Mathematical Olympiad

First Day

August 12, 2015

8:00–12:00

1. As shown in the following figure, in the acute triangle ABC, $AB > AC$; O is its circumcenter, and D is the midpoint of BC. The circle with diameter AD meets the sides AB and AC again at points E and F, respectively. The line passing through D parallel to AO meets EF at M. Show that $EM = MF$. (posed by Zheng Huan)

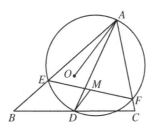

Fig. 1.1

Proof. As in Figure 1.1, connect DE, DF, and draw the line ON through O and perpendicular to AB that meets AB at N. By the assumptions,

97

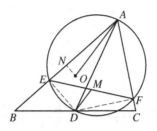

Fig. 1.2

$DE \perp AB, DF \perp AC$. So, $ON \parallel DE$. And since $DM \parallel AO$, so $\angle EDM = \angle AON$.

Since O is the circumcenter of $\triangle ABC$, $\angle AON = \angle ACB$. Therefore, $\angle EDM = \angle ACB$.

Similarly, $\angle FDM = \angle ABC$.

In $\triangle EDF$,

$$\frac{EM}{MF} = \frac{DE \cdot \sin \angle EDM}{DF \cdot \sin \angle FDM} = \frac{DE \cdot \sin \angle ACB}{DF \cdot \sin \angle ABC}$$

$$= \frac{DB \cdot \sin \angle ABC \cdot \sin \angle ACB}{DC \cdot \sin \angle ACB \cdot \sin \angle ABC} = 1.$$

So, $EM = MF$. □

2 Let $a \in (0,1)$ and

$$f(x) = ax^3 + (1 - 4a)x^2 + (5a - 1)x + (3 - 5a),$$

$$g(x) = (1 - a)x^3 - x^2 + (2 - a)x - (3a + 1).$$

Prove that, for any real number x, at least one of $|f(x)|$ and $|g(x)|$ is no less than $1 + a$. (posed by Li Shenghong)

Proof. Since $a \in (0, 1)$, both a and $1 - a$ are positive, so, for any real number x,

$$\max\{|f(x)|, |g(x)|\}$$

$$= (1 - a)\max\{|f(x)|, |g(x)|\} + a\max\{|f(x)|, |g(x)|\}$$

$$\geq (1 - a)|f(x)| + a|g(x)|$$

$$\geq |(1 - a)f(x) - ag(x)|.$$

We have,

$$(1-a)f(x) - ag(x)$$
$$= [(1-a)(1-4a)+a]x^2 + [(1-a)(5a-1)-a(2-a)]x$$
$$+ [(1-a)(3-5a)+a(3a+1)]$$
$$= (2a-1)^2(x^2-x+2)+1+a.$$

Since $x^2 - x + 2 = (x-1/2)^2 + 7/4 > 0$, so

$$\max\{|f(x)|, |g(x)|\} \geq 1+a. \qquad \square$$

3. In a grid of 12×12 unit squares, color each unit square with either black or white, such that there is at least one black unit square in any 3×4 or 4×3 rectangle bounded by the grid lines. Determine, with proof, the minimum number of black unit squares. (posed by Liang Yingde)

Solution The minimum number n is 12.

We first prove that $n \geq 12$. The 12×12 unit squares can be partitioned into 12 non-overlapping 3×4 rectangles. Since each such rectangle contains at least one black square, so there are at least 12 black squares.

On the other hand, we provide an example with 12 black squares that satisfy the requirement, as in Figure 3.1.

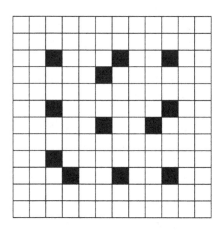

Fig. 3.1 $\qquad \square$

4. For any positive integer n, let $g(n)$ be the greatest common divisor of n and 2015. Determine the number of ordered triples (a, b, c) that satisfy the following conditions:

(1) $a, b, c, \in \{1, 2, \ldots, 2015\}$;

(2) $g(a), g(b), g(c), g(a+b), g(b+c), g(c+a), g(a+b+c)$ are pairwise different.

(posed by Wang Bin)

Solution   Factorize 2015 into primes, $2015 = 5 \times 13 \times 31$.

For any n, $g(n)$ is one of the 8 divisors of 2015.

We classify a number n as Type 0 if $g(n) = 1$; Type 1 if $g(n)$ is one of $5, 13, 31$; Type 2 if $g(n)$ is one of $5 \times 13, 13 \times 31, 31 \times 5$.

We will use two simple facts:

- For any integer x,

$$g(x) = g(-x) = g(2015 + x) = g(2015 - x).$$

This allows us to work in residue classes modulo 2015.

- For a prime p, if both $p \mid x$ and $p \mid y$, we have $p \mid x \pm y$; on the other hand, if exactly one of $p \mid x$ and $p \mid y$ holds, we have $p \nmid x \pm y$.

For any triple (a, b, c) satisfying the conditions in our problem statement, we write the tuple

$$A = (a, b, c, a+b, a+c, b+c, a+b+c),$$

and consider the g values on the seven positions in A. First we see that 2015 is not among the seven g values — for any x in A, there are y and z in other two positions where $x = y + z$ or $x = y - z$; $g(x) = 2015$ implies $g(y) = g(z)$. So, the seven g values must be $1, 5, 13, 31, 5 \times 13, 13 \times 31, 31 \times 5$, one for each. Thus A has one Type 0 number, three Type 1 numbers, and three Type 2 numbers. We will focus on the three positions for the Type 2 numbers.

Let p_1, p_2, p_3 be an arbitrary permutation of $5, 13, 31$.

If x, y, and z satisfy $g(x) = p_1 p_2$, $g(y) = p_1 p_3$, $g(z) = p_2 p_3$, then

$$p_1 \mid x, p_1 \mid y \Rightarrow p_1 \mid \pm x \pm y,$$

$$p_2 \mid x, p_2 \nmid y \Rightarrow p_2 \nmid \pm x \pm y,$$

$$p_3 \nmid x, p_3 \mid y \Rightarrow p_3 \nmid \pm x \pm y.$$

So, $g(\pm x \pm y) = p_1$. similarly, $g(\pm x \pm z) = p_2$, $g(\pm y \pm z) = p_3$, $g(\pm x \pm y \pm z) = 1$.

So, when we fix three positions in A for the three Type 2 numbers x, y, and z, if the other four numbers can be expressed as $\pm x \pm y$, $\pm x \pm z$, $\pm y \pm z$, and $\pm x \pm y \pm z$, we have different g values for all the seven positions. We call the set of these three positions *feasible configuration*.

In a feasible configuration, when we fix the values for x, y, and z, modulo 2015, then the number modulo 2015 on each position is uniquely determined. Under the requirement that $a, b, c \in \{1, 2, \ldots, 2015\}$, the integers a, b, c are uniquely determined. Modulo 2015, there are exactly $p_3 - 1$ values for x (satisfying $g(x) = p_1 p_2$), $p_2 - 1$ values for y, $p_1 - 1$ values for z. Recall that p_1, p_2, p_3 is any of the 6 permutations of $5, 13, 31$. So, in any feasible configuration, the number of possible (x, y, z) modulo 2015 is $6(5 - 1)(13 - 1)(31 - 1) = 8640$. These correspond to 8640 possible triples (a, b, c).

Triples for different feasible configurations must be different, because their corresponding A's are different.

In the following, by considering which positions have Type 2 numbers, we count the number of feasible configurations, as well as verify that no desired triples arise from non-feasible configurations.

For each prime p in $\{5, 13, 31\}$, exactly 3 positions in the tuple A is a multiple of p. If at least two of $a + b$, $b + c$, $c + a$ are multiples of p, without loss of generality, $p \mid a + b$ and $p \mid a + c$, then, under these assumption,

$$p \mid a \Leftrightarrow p \mid b \Leftrightarrow p \mid a + b + c,$$

$$p \mid b + c \Rightarrow p \mid (a + b) + (a + c) - (b + c) \Rightarrow p \mid 2a \Rightarrow p \mid a.$$

In any case, we would not have exactly three positions in A that are multiples of p. So, $5 \times 13 \times 31 \mid g(a + b) g(a + c) g(b + c)$; there are at most one Type 2 numbers among $a + b$, $b + c$, $c + a$. In other words, there are at least two Type 2 numbers among a, b, c, and $a + b + c$.

Case 1. There are three Type 2 numbers among a, b, c, and $a + b + c$. We consider two sub-cases

Case 1.1. $a + b + c$ is not in Type 2. $x = a$, $y = b$, $z = c$ are Type 2 numbers. Then, $a + b = x + y$, $b + c = y + z$, $a + c = x + z$, $a + b + c = x + y + z$, this is a feasible configuration.

Case 1.2. $z = a + b + c$ is in Type 2. First assume the other Type 2 numbers are $x = a$, $y = b$. Then, $a + b = x + y$, $b + c = -x + z$, $a + c = -y + z$, and $c = -x - y + z$; we get a feasible configuration. Similarly, when the other Type 2 numbers are b and c, or a and c, we get other two feasible configurations.

Case 2. There are exactly two Type 2 numbers among a, b, c, and $a + b + c$. We consider two sub-cases.

Case 2.1. $a + b + c$ is not in Type 2. Without loss of generality, a and b are two type two numbers. Then, it is easy to see $a + b$ cannot be the other Type 2 number. We first assume $x = a$, $y = b$, $z = a + c$ are the three Type 2 numbers. Now, $a + b = x + y$, $c = -x + z$, $a + b + c = y + z$, $b + c = -x + y + z$. By symmetry, we have 6 feasible configurations in this sub-case.

Case 2.2. $a + b + c$ is a Type 2 number. Now we assume a is another Type 2 number among a, b, and c. It is easy to see that $b + c$ can not be a Type 2 number, one of $a + b$ and $a + c$ is. By symmetry, we assume $x = a$, $y = a + b + c$, $z = a + b$ are Type 2 numbers. Now $b + c = -x + y$, $b = -x + z$, $c = y - z$, $a + c = x + y - z$. By symmetry, we get 6 feasible configurations in this sub-case.

To conclude, we have 16 different feasible configurations. So the number of desired triples is $16 \times 8640 = 138240$. $\square$

Second Day

August 13, 2015

8:00–12:00

5   Determine the number of distinct right-angled triangles such that its three sides are of integral lengths, and its area is 999 times of its perimeter. (Congruent triangles are considered identical.) (posed by Lin Chang)

Solution (First solution). Suppose the three sides are $a < b < c$. Let r be the inradius of the triangle, so $r = (a + b - c)/2$. By Pythagoras, $a^2 + b^2 = c^2$. Under these assumptions, for any positive integer m,

$$S = \frac{1}{2}r(a + b + c) = m(a + b + c)$$

$$\Leftrightarrow r = (a + b - c)/2 = 2m$$

$$\Leftrightarrow c = a + b - 4m$$

$$\Leftrightarrow ab - 4ma - 4mb + 8m^2 = 0, 4m < a < b$$

$$\times \text{ (by Pythagoras and } a < b < c)$$

$$\Leftrightarrow (a - 4m)(b - 4m) = 8m^2, a - 4m > 0, b - 4m > 0.$$

Any solution pair (a, b) with $a < b$, $a - 4m > 0$, and $b - 4m > 0$ to the last equation gives a different triangle. Clearly no solution has $a = b$ because $8m^2$ is not a square. Therefore, the number of such triangles is $d(8m^2)/2$, where $d(n)$ is the number of divisors of a positive integer n.

In our problem, $m = 999$, $8m^2 = 2^3 \times 3^6 \times 37^2$, the answer is $d(8m^2)/2 = (3 + 1)(6 + 1)(2 + 1)/2 = 42$.   □

Solution (Second solution). It is well known that the Pythagorean numbers are in the form

$$a = k \cdot 2uv, \quad b = k(u^2 - v^2), \quad c = k(u^2 + v^2),$$

where k — the greatest common divisor of three sides — can be any positive integer, u and v are coprime, are of different parity, and $u > v$.

In our problem,

$$\frac{1}{2}ab = 999(a + b + c)$$

$$\Leftrightarrow k^2 uv(u^2 - v^2) = 999 \times 2ku(u + v)$$

$$\Leftrightarrow kv(u - v) = 1998 = 2 \times 3^3 \times 37.$$

$u - v$ is an odd number, the factor 2 can go to k or v. v and $u - v$ are coprime, any odd prime factor p^α distributed to k, v, $u - v$ must be in the form $(\alpha, 0, 0)$, $(i, \alpha - i, 0)$, or $(i, 0, \alpha - i)$ $(1 \le i \le \alpha)$, we have $2\alpha + 1$ choices here.

By the multiplication rule in counting, we have $2 \times (2 \times 3 + 1) \times (2 \times 1 + 1) = 42$ ways to distribute the prime factors, each of which corresponds to a unique triangle. Therefore, the answer is 42.

More generally, if 2015 is replaced by any number $m = 2^\alpha p_1^{\beta_1} \ldots p_k^{\beta_k}$, where $p_1, p_2, \ldots, p_k$ are distinct odd primes, the number of triangles satisfying the condition is $(\alpha + 2)(2\beta_1 + 1) \ldots (2\beta_k + 1)$. $\qquad \square$

6 As shown in the figure, Γ_1 and Γ_2 are two circles lying outside each other. AB is an external tangent line to Γ_1 and Γ_2, with points of tangency A and B, respectively. CD is an internal tangent line, with points of tangency C and D, respectively. Lines AC and BD meets at E. F is a point on Γ_1. The line passing through F and tangent to Γ_1 meets the perpendicular bisector of EF at M. MG is a line tangent to Γ_2 at point G. Prove that $MF = MG$. (posed by Fu Yunhao)

Proof (First proof). As in Figure 6.1, let O_1 and O_2 be the centers of Γ_1 and Γ_2, respectively; let H be the intersection of the lines AB and CD; connect HO_1 and HO_2.

Let J and K be the midpoints of AB and CD, respectively; connect JE, JK.

Because HA and HC are tangent to Γ_1, HO_1 bisects $\angle AHC$, and $AC \perp HO_1$. Similarly, HO_2 bisects $\angle BHD$ and $BD \perp HO_2$. HO_1 and HO_2 are the angle bisector and the exterior angle bisector of $\angle AHC$, so $HO_1 \perp HO_2$. Since $AC \perp HO_1$ and $BD \perp HO_2$, so $AC \perp BD$.

In the right triangle AEB, J is the midpoint of the hypotenuse, so $JE = JA = JB$; similarly, $KE = KC = KD$. Consider Γ_1, Γ_2, and the circle with center E and radius 0. Since $JE = JA = JB$, J has the same power to the three circles. Similarly, K has the same power to the

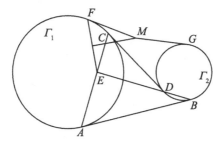

Fig. 6.1

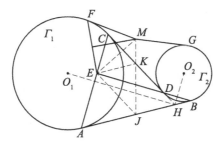

Fig. 6.2

three circles. Clearly J and K are two distinct points, so these three circles has a common radical axis. M is on the perpendicular bisector of EF, so $ME = MF$; and MF is the tangent line to Γ_1, so M is on the common radical axis of the three circles. Now MG is tangent to Γ_2, so $MF = MG$.

Proof (Second proof). We use the same argument to get $AC \perp BD$. Let r_1 and r_2 be the radius of Γ_1 and Γ_2, respectively. By Pythagoras theorem, $JO_1^2 - JE^2 = r_1^2 + JA^2 - JE^2 = r_1^2$. Similarly, $KO_1^2 - KE^2 = r_1^2$, $MO_1^2 - ME^2 = r_1^2$. So,

$$JO_1^2 - JE^2 = KO_1^2 - KE^2 = MO_1^2 - ME^2.$$

It is easy to see that $JK \perp O_1E$, $KM \perp O_1E$, thus J, K, M are collinear. $JO_2^2 - JE^2 = r_2^2 + JB^2 - JE^2 = r_2^2$. Similarly, $KO_2^2 - KE^2 = r_2^2$. So $JO_2^2 - JE^2 = KO_2^2 - KE^2$, therefore $JK \perp O_2E$. So, $JM \perp O_2E$. This implies $MO_2^2 - ME^2 = JO_2^2 - JE^2 = r_2^2$. Note that $MO_2^2 = MG^2 + r_2^2$, so $MG = ME$, and $MG = MF$. □

Note: The point E is on the line O_1O_2. This fact can be inferred from both proofs.

7 Suppose $n \geq 2$ and $x_1, x_2, \ldots x_n \in (0, 1)$. Prove that

$$\frac{\sqrt{1 - x_1}}{x_1} + \frac{\sqrt{1 - x_2}}{x_2} + \cdots + \frac{\sqrt{1 - x_n}}{x_n} < \frac{\sqrt{n - 1}}{x_1 x_2 \ldots x_n}.$$

(posed by Wang Xinmao)

Proof (First proof). We prove by induction on n. When $n = 2$, by the Cauchy-Schwarz inequality,

$$\frac{\sqrt{1 - x_1}}{x_1} + \frac{\sqrt{1 - x_2}}{x_2}$$
$$= \frac{x_2\sqrt{1 - x_1} + x_1\sqrt{1 - x_2}}{x_1 x_2}$$
$$\leq \frac{\sqrt{1 - x_1 + x_1^2}\sqrt{1 - x_2 + x_2^2}}{x_1 x_2}$$
$$< \frac{1}{x_1 x_2}.$$

When $n \geq 3$, by the inductive hypothesis and the Cauchy-Schwarz inequality,

$$\frac{\sqrt{1 - x_1}}{x_1} + \frac{\sqrt{1 - x_2}}{x_2} + \cdots + \frac{\sqrt{1 - x_n}}{x_n}$$
$$< \frac{\sqrt{n - 2}}{x_1 x_2 \ldots x_{n-1}} + \frac{\sqrt{1 - x_n}}{x_n}$$
$$= \frac{\sqrt{n - 2}x_n + x_1 x_2 \ldots x_{n-1}\sqrt{1 - x_n}}{x_1 x_2 \ldots x_n}$$
$$\leq \frac{\sqrt{n - 2 + (x_1 x_2 \ldots x_{n-1})^2}\sqrt{1 - x_n + x_n^2}}{x_1 x_2 \ldots x_n}$$
$$< \frac{\sqrt{n - 1}}{x_1 x_2 \ldots x_n}.$$

Proof (Second proof). Let $A = x_1x_2\ldots x_n(x_1^{-1} + x_2^{-1} + \cdots + x_n^{-1})$, and $B = x_1x_2\ldots x_n$. Multiply both sides of the inequality to be proved by B, we need to show that

$$\sqrt{1 - x_1}x_2x_3\ldots x_n + \sqrt{1 - x_2}x_3x_4\ldots x_nx_1 + \ldots$$
$$+ \sqrt{1 - x_n}x_2x_3\ldots x_n < \sqrt{n - 1}.$$

By the Cauchy-Schwarz inequality,

the left hand side of the above equation

$$\leq \sqrt{x_2\ldots x_n + x_3x_4\ldots x_nx_1 + \ldots x_1x_2\ldots x_{n-1}} \cdot$$

$$\sqrt{(1 - x_1)x_2x_3\ldots x_n + (1 - x_2)x_3x_4\ldots x_nx_1 + \cdots + (1 - x_n)x_1x_2\ldots x_n}$$

$$= \sqrt{A(A - nB)}.$$

So we only need to show $A(A - nB) < n - 1$.

We first show that $A < 1 + (n - 1)B$. [1] Indeed,

$$1 + (n - 1)B - A = (1 - x_1)(1 - x_2x_3\ldots x_n)$$
$$+ x_1(1 - x_2)(1 - x_3x_4\ldots x_n)$$
$$+ x_1x_2(1 - x_3)(1 - x_4x_5\ldots x_n) + \ldots$$
$$+ x_1x_2\ldots x_{n-2}(1 - x_{n-1})(1 - x_n) > 0.$$

Note that $B < 1$; so,

$$A(A - nB) < (1 + (n - 1)B)(1 + (n - 1)B - nB)$$
$$= (1 + (n - 1)B)(1 - B)$$
$$= -(n - 1)B^2 + (n - 2)B + 1$$
$$< (n - 2)B + 1 < (n - 2) + 1 = n - 1. \qquad \square$$

8 Let $n \geq 2$ be a given integer. Initially, we write n sets on the blackboard, and do a sequence of moves as follows: in each move, we choose two sets A and B on the blackboard such that none of them

---

[1] This can also be proved by mollification, or finding the extreme values of a linear function.

is a subset of the other, then replace A and B by the two sets $A \cup B$ and $A \cap B$. Find the maximum number of moves over all possible initial sets and all possible sequence of moves. (posed by Zhu Huawei)

Solution   We first prove that any such sequence has at most $\binom{n}{2}$ moves.

We call an (unordered) pair of sets A and B on the blackboard a *good pair* if one of them is a subset of the other. We prove that the number of good pairs increases at least by 1 after each move. Suppose we replace A and B with $A \cup B$ and $A \cap B$ in a move. By our rule, A and B do not form a good pair before the move, yet $A \cup B$ and $A \cap B$ form a good pair after the move. Consider any other set C on the blackboard.

(1) If C forms a good pair with exactly one of A and B, by symmetry, we assume $A \subseteq C$, then $A \cap B \subseteq C$, i.e. after the move C forms a good pair with at least one of the two new sets.

(2) C forms good pairs with both A and B. Since A and B is not a good pair, so $A \subseteq C$ and $B \subseteq C$, or else $C \subseteq A$ and $C \subseteq B$. In the former case, $A \cap B \subseteq C$ and $A \cup B \subseteq C$; in the latter, $C \subseteq A \cap B$ and $C \subseteq A \cup B$. So, the number of good pairs occur on C does not decrease after a move.

Hence, in each move, the number of good pairs increases by at least 1. The number is at least 0 and at most $\binom{n}{2}$. So there can be at most $\binom{n}{2}$ moves.

Next, we give an example with exactly $\binom{n}{2}$ moves.

Define the sets

$$A_i = \{i, i+1, \ldots, i+n-2\}, i = 1, 2, \ldots, n.$$

We prove by induction that, from these n sets, we have a sequence of $\binom{n}{2}$ moves.

When $n = 2$, $A_1 = \{1\}$ and $A_2 = \{2\}$, and we can have $\binom{2}{2} = 1$ moves.

Suppose the proposition holds for n, now consider the $n + 1$ sets. Our first move involves $A_1 = \{1, 2, \ldots, n\}$ and $A_2 = \{2, 3, \ldots, n+1\}$, and they are replaced by $\{2, 3, \ldots, n\}$ and $\{1, 2, \ldots, n+1\}$; the next move involves the latter set and $A_3 = \{3, 4, \ldots, n+2\}$, we get $\{3, 4, \ldots, n+1\}$ and $\{1, 2, \ldots, n+2\}$; the next move involves the latter set and A_4; and so on.

After n moves, we get one set $\{1, 2, \ldots, 2n\}$ that is the union of all the initial sets. The other n sets are $\{2, 3, \ldots, n\}$, $\{3, 4, \ldots, n+1\}$, $\ldots$, $\{n+1, n+2, \ldots, 2n-1\}$. The other moves are among these n sets.

Clearly, decreasing each element by 1 does not affect the inclusion relations between sets. So those n sets are equivalent to $\{1, 2, \ldots, n-1\}$, $\{2, 3, \ldots, n\}$, $\ldots$, $\{n, n+1, \ldots, 2n-2\}$. By inductive hypothesis, we have $\binom{n}{2}$ moves from these sets. So, for the initial $n+1$ sets, the number of moves can be

$$n + \binom{n}{2} = \binom{n+1}{2}.$$

In conclusion, the maximum number of moves is $\binom{n}{2} = n(n-1)/2$. $\quad\square$

China Western Mathematical Olympiad

First Day

August 16, 2014

8:00–12:00

1. For given positive reals x and y, find the minimum value of
$$x + y + \frac{|x-1|}{y} + \frac{|y-1|}{x}.$$
(posed by Gu Bin and Cen Aiguo)

Solution Let
$$f(x,y) = x + y + \frac{|x-1|}{y} + \frac{|y-1|}{x}.$$
When $x \geq 1$ and $y \geq 1$, $f(x,y) \geq x + y \geq 2$.

When $0 < x \leq 1$ and $0 < y \leq 1$,
$$f(x,y) = x + y + \frac{1-x}{y} + \frac{1-y}{x} \geq x + y + 1 - x + 1 - y = 2.$$

Otherwise, without loss of generality, $0 < x < 1 < y$,
$$f(x,y) = x + y + \frac{1-x}{y} + \frac{y-1}{x}$$
$$= y + \frac{1}{y} + \frac{xy-x}{y} + \frac{y-1}{x}$$

$$= y + \frac{1}{y} + (y-1)\left(\frac{x}{y} + \frac{1}{x}\right)$$

$$\geq 2\sqrt{y \cdot \frac{1}{y}} + 0$$

$$= 2.$$

So, $f(x,y) \geq 2$ for any $x > 0$ and $y > 0$.

$f(1,1) = 2$. So the answer is 2.   □

2   As shown in Figure 2.1, AB is a diameter of a semicircle with center O, points C and D are on the arc $\overset{\frown}{AB}$, points P and Q are the circumcenters of $\triangle OAC$ and $\triangle OBD$, respectively. Prove that $CP \cdot CQ = DP \cdot DQ$. (posed by He Yijie)

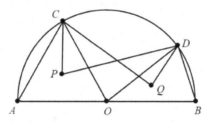

Fig. 2.1

Proof. As in Figure 2.2, connect OP, OQ, AP, AD, BQ, BC. Assume $\angle BAD = \alpha$, and $\angle ABC = \beta$.

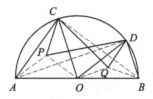

Fig. 2.2

We have,

$$\angle OAP = \angle AOP = \frac{1}{2}\angle AOC = \angle ABC = \beta,$$

$$\angle OBQ = \angle BOQ = \frac{1}{2}\angle BOD = \angle BAD = \alpha.$$

So,

$$\angle PAD = \angle OAP - \angle OAD = \beta - \alpha = \angle OBC - \angle OBQ = \angle QBC.$$

Also,

$$\frac{AD}{AP} = \frac{AD}{AB}\frac{AB}{AO}\frac{AO}{AP} = \cos\alpha \cdot 2 \cdot 2\cos\beta = 4\cos\alpha\cos\beta.$$

Similarly, $\dfrac{BC}{BQ} = 4\cos\alpha\cos\beta$; so $\dfrac{AD}{AP} = \dfrac{BC}{BQ}$, and $\triangle APD \sim \triangle BQC$. So, $\dfrac{AP}{DP} = \dfrac{BQ}{CQ}$, and

$$CP \cdot CQ = AP \cdot CQ = DP \cdot BQ = DP \cdot DQ. \qquad \square$$

3 Let $A_1, A_2, A_3, \ldots$ be a sequence of sets satisfying the following condition: for any positive integer j, there exist only finitely many positive integers i such that $A_i \subseteq A_j$. Prove that there exists a sequence $a_1, a_2, a_3, \ldots$ of positive integers such that, for any positive integers i and j, $a_i \mid a_j$ if and only if $A_i \subseteq A_j$. (posed by Qu Zhenhua)

Proof. List all the primes in increasing order as $p_1, p_2, p_3, \ldots$. and for any $i \in \mathbb{N}^*$, define

$$S_i = \{j \in \mathbb{N}^* : A_j \subseteq A_i\}.$$

By our assumption, S_i is finite for all i, and is not empty since $i \in S_i$. Let

$$a_i = \prod_{j \in S_i} p_j.$$

We prove that $a_1, a_2, a_3, \ldots$ is the desired sequence.

For any integers i and j, when $A_i \subseteq A_j$, we have $S_i \subseteq S_j$, therefore $a_i \mid a_j$; when $a_i \mid a_j$, then $S_i \subseteq S_j$, and since $i \in S_i$, so $i \in S_j$, thus $A_i \subseteq A_j$. So, $a_i \mid a_j$ if and only if $A_i \subseteq A_j$. $\qquad \square$

4 Given a positive integer n, let $A = (a_1, a_2, \ldots, a_n)$ be a sequence of non-negative integers. A sub-sequence of one or more consecutive terms of A is called a *dragon*, if its arithmetic mean is no less than 1; in this case, the first term and the last term of the sub-sequence are called a *dragon-head* and a *dragon-tail*, respectively. Suppose that every term of A is either a dragon-head or a dragon-tail, determine the smallest possible value of $\sum_{i=1}^{n} a_i$. (posed by Zou Jin)

Solution　We prove that the smallest possible value of $\sum_{i=1}^{n} a_i$ is $\lfloor n/2 \rfloor + 1$.

First, we give the examples where the minimum is achieved. When $n = 2k - 1$ is an odd number, let $a_k = k$ and all the other terms be 0; and when $n = 2k$ is even, let $a_k = k$, $a_{2k} = 1$, and all the other terms 0. It is easy to verify that each term is a dragon-head or dragon-tail, and $\sum_{i=1}^{n} a_i = \lfloor n/2 \rfloor + 1$.

Now we prove by induction that any series $a_1, a_2, \ldots, a_n$ satisfying the condition has $\sum_{i=1}^{n} a_i \geq \lfloor n/2 \rfloor + 1$.

When $n = 1$, the proposition clearly holds.

Suppose the proposition holds for any series with less than n terms, and $a_1, a_2, \ldots, a_n$ is a series where every term is a dragon-head or dragon-tail.

Let t be the length of the longest dragon with a_1 as its dragon-head, if $t \geq \lfloor n/2 \rfloor + 1$, we are done.

Otherwise, $t \leq \lfloor n/2 \rfloor$. By the definition of a dragon and the maximality of t, we have $a_1 + a_2 + \cdots + a_t = t$ and $a_{t+1} = 0$. Let $b_1 = a_{t+1} + a_{t+2} + \cdots + a_{2t}$, $b_2 = a_{2t+1}$, $b_3 = a_{2t+2}$, $\ldots$, $b_{n-2t+1} = a_n$.

Now we prove that each term in $b_1, b_2, \ldots, b_{n-2t+1}$ is a dragon-head or a dragon-tail.

If a_{i-2t-1} is a dragon-head, so is b_i.

If a_{i-2t-1} is a dragon-tail, then there exists positive integer m such that

$$a_m + a_{m+1} + \cdots + a_{i+2t-1} \geq i + 2t - m.$$

Case 1. $m \geq 2t + 1$. We have

$$b_{m-2t+1} + b_{m-2t+2} + \cdots + b_i = a_m + a_{m+1} + a_{i+2t-1} \geq i + 2t - m.$$

Case 2. $t + 1 \leq m \leq 2t$. We have

$$b_1 + b_2 + \cdots + b_i \geq a_m + a_{m+1} + \cdots + a_{i+2t-1} \geq i + 2t - m \geq i.$$

Case 3. $m \leq t$. We have

$$b_1 + b_2 + \cdots + b_i = a_1 + a_2 + \cdots + a_{i+2t-1} - t \geq i + 2t - m - t \geq i.$$

In any case, b_i is a dragon-tail.

Now, by the inductive hypothesis,

$$\sum_{i=1}^{n-2t+1} b_i \geq \left\lfloor \frac{n - 2t + 1}{2} \right\rfloor + 1.$$

So,

$$\sum_{i=1}^{n} a_i = t + \sum_{i=1}^{n-2t+1} b_i \geq t + \left\lfloor \frac{n-2t+1}{2} \right\rfloor + 1 \geq \left\lfloor \frac{n}{2} \right\rfloor + 1.$$

In conclusion, the minimum value of $\sum_{i=1}^{n} a_i$ is $\lfloor n/2 \rfloor + 1$. □

Second Day

August 17, 2014

8:00–12:00

5 Given a positive integer m, prove that there exists a positive integer n_0 such that the first digits after the decimal point in the decimal representation of $\sqrt{n^2 + 817n + m}$ are the same for all integers $n > n_0$. (posed by Feng Zhigang)

Proof.

$$\sqrt{n^2 + 817n + m} = \sqrt{\left(n + \frac{817}{2}\right)^2 + m - \frac{817^2}{4}}.$$

(1) When $m > \dfrac{817^2}{4}$, let n_0 be a positive integer bigger than $5m - 5\left(408 + \dfrac{3}{5}\right)^2$. Then, for any $n > n_0$,

$$\left(n + \frac{817}{2}\right)^2 < \left(n + \frac{817}{2}\right)^2 + m - \frac{817^2}{4} < \left(n + 408 + \frac{3}{5}\right)^2,$$

and the first digit of $\sqrt{n^2 + 817n + m}$ after the decimal point is 5.

(2) When $m < \dfrac{817^2}{4}$, let n_0 be a positive integer bigger than $5\left(408 + \dfrac{2}{5}\right)^2 - 5m$. Then, for any $n > n_0$,

$$\left(n + 408 + \frac{3}{5}\right)^2 < \left(n + \frac{817}{2}\right)^2 + m - \frac{817^2}{4} < \left(n + \frac{817}{2}\right)^2,$$

and the first digit of $\sqrt{n^2 + 817n + m}$ after the decimal point is 4.

So, there is always a positive integer n_0, such that the first digits after the decimal point in the decimal representation of $\sqrt{n^2 + 817n + m}$ are the same for all integers $n > n_0$. $\qquad\square$

6 Given an integer $n \geq 2$, let $x_1, x_2, \ldots, x_n$ be real numbers satisfying the following conditions:

(1) $\sum_{i=1}^{n} x_i = 0$;

(2) $|x_i| \leq 1, i = 1, 2, \ldots, n$.

Determine the maximum value of $\min_{1 \le i \le n-1} |x_i - x_{i+1}|$. (posed by Leng Gangsong)

Solution Let $A = \min_{1 \le i \le n-1} |x_i - x_{i+1}|$. We prove that the maximum value of A is 2 when n is even, and $\dfrac{2n}{n+1}$ when n is odd.

(a) When n is even, by (2), $|x_i - x_{i+1}| \le |x_i| + |x_{i+1}| \le 2$ for all $1 \le i \le n-1$. So, $A \le 2$. The sequence defined by $x_i = (-1)^i$, $i = 1, 2, \ldots, n$, satisfies (1) and (2), and $A = 2$. So the maximum value of A is 2.

(b) When n is odd, $n = 2k + 1$.

If $x_i \le x_{i+1} \le x_{i+2}$ or $x_i \ge x_{i+1} \ge x_{i+2}$ for some i, we have $2A \le |x_{i+2} - x_i| \le 2$ since $A \le |x_i - x_{i+1}|$ and $A \le |x_{i+1} - x_{i+2}|$; therefore, $A \le 1 < \dfrac{2n}{n+1}$.

Otherwise, without loss of generality, $x_{2i-1} > x_{2i}$ and $x_{2i} < x_{2i+1}$ for $i = 1, 2, \ldots, k$. Then,

$$(2k+2)A \le \sum_{i=1}^{k} (|x_{2i-1} - x_{2i}| + |x_{2i} - x_{2i+1}|) + (|x_1 - x_2| + |x_{2k} - x_{2k+1}|)$$

$$= \sum_{i=1}^{k} (x_{2i-1} - x_{2i} + x_{2i+1} - x_{2i}) + (x_1 - x_2 + x_{2k+1} - x_{2k})$$

$$= 2 \sum_{i=1}^{k+1} x_{2i-1} - 2 \sum_{i=1}^{k} x_{2i} - x_2 - x_{2k}$$

$$= -4 \sum_{i=1}^{k} x_{2i} - x_2 - x_{2k}$$

$$\le 4k + 2.$$

So, $A \le \dfrac{4k+2}{2k+2} = \dfrac{2n}{n+1}$. The sequence defined by

$$x_i = \begin{cases} \dfrac{k}{k+1}, & i = 1, 3, \ldots, 2k+1, \\ -1, & i = 2, 4, \ldots, 2k, \end{cases}$$

satisfies (1) and (2); and it is easy to verify that, for this sequence, $A = \dfrac{2k+1}{k+1} = \dfrac{2n}{n+1}$. $\qquad \square$

7 Let O be the center of an equilateral triangle ABC in a plane, and P, Q be two points such that $\overrightarrow{OQ} = 2\overrightarrow{PO}$. Prove that

$$|PA| + |PB| + |PC| \leq |QA| + |QB| + |QC|.$$

(posed by Yang Mingliang)

Proof (First proof). Let A_1, B_1, and C_1 be the midpoints of BC, CA, and AB, respectively.

The homothety with center O and scale factor -2 sends $\triangle A_1 B_1 C_1$ to $\triangle ABC$, and sends P to Q. So,

$$QA + QB + QC = 2(PA_1 + PB_1 + PC_1).$$

In the quadrilateral $PA_1 B C_1$, by Ptolemy's inequality,

$$PB \cdot A_1 C_1 \leq PC_1 \cdot A_1 B + PA_1 \cdot BC_1.$$

Note that $\triangle A_1 B C_1$ is equilateral, so $PB \leq PA_1 + PC_1$. Similarly, $PC \leq PA_1 + PB_1$, and $PA \leq PB_1 + PC_1$. Adding the three inequalities together, we have

$$PA + PB + PC \leq 2(PA_1 + PB_1 + PC_1).$$

So, $PA + PB + PC \leq QA + QB + QC$. $\qquad\square$

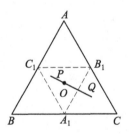

Fig. 7.1

Proof (Second proof). We first show that $QA + QB \geq 2PC$.

Set $\omega = e^{i2\pi/3}$. $\overrightarrow{OB} = \omega \overrightarrow{OA}$, $\overrightarrow{OC} = \omega^2 \overrightarrow{OA}$, $\overrightarrow{OQ} = -2\overrightarrow{OP}$, so,

$$QA + QB \geq 2PC$$
$$\Leftrightarrow |\overrightarrow{OA} - \overrightarrow{OQ}| + |\overrightarrow{OB} - \overrightarrow{OQ}| \geq 2|\overrightarrow{OC} - \overrightarrow{OP}|$$

$$\Leftrightarrow |\overrightarrow{OA} + 2\overrightarrow{OP}| + |\omega\overrightarrow{OA} + 2\overrightarrow{OP}| \geq 2|\omega^2\overrightarrow{OA} - \overrightarrow{OP}|$$

$$\Leftrightarrow |\overrightarrow{OA} + 2\overrightarrow{OP}| + |\overrightarrow{OA} + 2\omega^2\overrightarrow{OP}| \geq 2|\overrightarrow{OA} - \omega\overrightarrow{OP}|.$$

Indeed, by the triangle inequality,

$$|\overrightarrow{OA} + 2\overrightarrow{OP}| + |\overrightarrow{OA} + 2\omega^2\overrightarrow{OP}|$$

$$\geq |\overrightarrow{OA} + 2\overrightarrow{OP} + \overrightarrow{OA} + 2\omega^2\overrightarrow{OP}|$$

$$= 2|\overrightarrow{OA} + (1 + \omega^2)\overrightarrow{OP}|$$

$$= 2|\overrightarrow{OA} - \omega\overrightarrow{OP}|,$$

so $QA + QB \geq 2PC$. Similarly, $QB + QC \geq 2PA$, and $QC + QA \geq 2PB$. Adding the three together, we have

$$PA + PB + PC \leq QA + QB + QC. \qquad \square$$

8. Given a real number q, $1 < q < 2$, define a sequence $\{x_n\}$ as follows: for any positive integer n, let

$$n = a_0 + a_1 \cdot 2 + a_2 \cdot 2^2 + \cdots + a_k \cdot 2^k, \quad (a_i \in \{0,1\}, i = 0, 1, \ldots, k),$$

be its binary representation, define

$$x_n = a_0 + a_1 \cdot q + a_2 \cdot q^2 + \cdots + a_k \cdot q^k.$$

Prove that for any positive integer n, there exists a positive integer m such that $x_n < x_m \leq x_n + 1$. (posed by Chen Yonggao)

Proof. Since $x_{2^k} = q^k$, we see that the sequence $\{x_n\}$ is not bounded from above.

For any positive integer n, define m to be the minimum positive integer such that $x_m > x_n$, we prove that $x_m \leq x_n + 1$.

Clearly $m > 1$. We divide the discussion into two cases.

Case 1. $m - 1$ is even. Suppose $m - 1 = a_1 \cdot 2^1 + \cdots + a_k \cdot 2^k$, $a_i \in \{0,1\}$ for each $1 \leq i \leq k$. Then, $m = 1 + a_1 \cdot 2^1 + \cdots + a_k \cdot 2^k$, and

$$x_m = 1 + a_1 \cdot q^1 + \cdots + a_k \cdot q^k = x_{m-1} + 1.$$

By the definition of m, $x_{m-1} \leq x_n$, so $x_m \leq x_n + 1$.

Case 2. $m - 1$ is odd. Let

$$m - 1 = 1 + 2 + \cdots + 2^l + a_{l+2} \cdot 2^{l+2} + \cdots + a_k \cdot 2^k,$$

where $l \geq 0$, $a_i \in \{0, 1\}$ for each $l + 2 \leq i \leq k$. Then,

$$
\begin{aligned}
x_m - x_{m-1} &= q^{l+1} - (1 + q + q^2 + \cdots + q^l) \\
&= (q-1)(1 + q + q^2 + \cdots + q^l) - (1 + q + q^2 + \cdots + q^l) + 1 \\
&< 1.
\end{aligned}
$$

Again, by the definition of m, $x_{m-1} \leq x_n$, so $x_m < x_{m-1} + 1 \leq x_n + 1$. In conclusion, we always have $x_m \leq x_n + 1$. $\square$

China Western
Mathematical Olympiad

2015 (Yin Chuan, Ning Xia)

August 16–17, 2015

1. Let n be a given positive integer, and $x_1, x_2, \ldots, x_n$ are real numbers such that $\sum_{i=1}^{n} x_i$ is an integer. Let $d_k = \min_{m \in \mathbb{Z}} |x_k - m|$, $1 \leq k \leq n$. Determine, with proof, maximum value of the sum $\sum_{i=1}^{n} x_i$.

Proof (First proof). WLOG, we may assume that all x_i's are in $(0, 1]$; otherwise, replace x_i by $x_i - [x_i]$. Let $t = \sum_{i=1}^{n} t$, it follows from $x_i \in [0, 1)$ that the integer $t \in [0, n)$. After rearranging the order, we may assume that $x_1, x_2, \ldots, x_k \leq \dfrac{1}{2}$, $x_{k+1}, x_{k+2}, \ldots, x_n > \dfrac{1}{2}$. It follows that $x_1 + x_2 + \cdots + x_k \leq \dfrac{k}{2}$ and $x_1 + x_2 + \cdots + x_k = t - (x_{k+1} + \cdots + x_n) \leq t - \dfrac{n-k}{2}$. Then

$$\sum_{k=1}^{n} d_k = x_1 + x_2 + \cdots + x_{k+1} - x_{k+1}, \cdots + 1 - x_n$$

$$= 2(x_1 + x^2 + \cdots + x_k) + n - k - t$$

$$= \min\{n-, t, t\} \leq \left[\dfrac{n}{2}\right].$$

The equality holds if we choose $x_1 = x_2 = \cdots = x_{n-1} = \dfrac{1}{2}$, and $x_n = \dfrac{1}{2}$ or 0 when n is even or odd respectively. One can check that the integral sum condition holds, and the sum $\sum_{i=1}^{n} d_i = \left[\dfrac{n}{2}\right]$. Hence, the maximum value is $\left[\dfrac{n}{2}\right]$.

Proof (Second proof). By replacing x_i by $x_i - [x_i]$ if necessary, we may assume $0 \le x_i < 1$ for all i. First note that $d_i = \min\{x_i, 1 - x_i\}$, then $\sum_{i=1}^{n} d_i \le \sum_{i=1}^{n} \min\{x_i, 1 - x_i\} \le \sum_{i=1}^{n} x_i$, and similarly, $\sum_{i=1}^{n} d_i \le \sum_{i=1}^{n}(1 - x_i)$. The given condition implies that both sums $\sum_{i=1}^{n} x_i$ and $\sum_{i=1}^{n}(1 - x_i)$ are integers. As these two sums adds up to n, so one of the sums is at most $\left[\dfrac{n}{2}\right]$, and hence $\sum_{i=1}^{n} d_i \le \left[\dfrac{n}{2}\right]$.

The equality holds if $x_1 = x_2 = \cdots = x_{n-1} = \dfrac{1}{2}$, and $x_n = \dfrac{1}{2}$ or 0 when n is even or odd, respectively. Hence, the maximum value is $\left[\dfrac{n}{2}\right]$.

Proof (Third proof). The given condition implies the global condition $\sum_i \{x_i\} = \sum_i (x_i - [x_i]) = (\sum_i x_i) - (\sum_i [x_i])$ is an integer, denoted by N. First note that $\{x_i\} \in [0, 1)$. If $\{x_i\} \le \dfrac{1}{2}$, then by definition we have $d_i = \{x_i\}$; otherwise, $x_i > \dfrac{1}{2}$, so $d_i = 1 - \{x_i\}$. It follows that

$$d_i + \left|\dfrac{1}{2} - \{x_i\}\right| = \dfrac{1}{2}, \text{ and the sum } \sum_i d_i = \dfrac{n}{2} - \sum_i \left|\dfrac{1}{2} - \{x_i\}\right| \le \dfrac{n}{2} - \left|\sum_i \dfrac{1}{2} - \{x_i\}\right| \le \dfrac{n}{2} - \left|\dfrac{n}{2} - N\right| = \dfrac{n}{2} - \left|\left[\dfrac{n}{2}\right] - N + \left\{\dfrac{n}{2}\right\}\right| \le \dfrac{n}{2} - \left\{\dfrac{n}{2}\right\} = \left[\dfrac{n}{2}\right].$$

The equality holds if $x_1 = x_2 = \cdots = x_{n-1} = \dfrac{1}{2}$, and $x_n = \dfrac{1}{2}$ or 0 when n is even or odd, respectively. Hence, the maximum value is $\left[\dfrac{n}{2}\right]$. $\qquad\square$

2 As shown in Figure 2.1, circles ω_1 and ω_2 are tangent to each other at the point T. M and N are two distinct points on ω_1 and different from T. AB and CD are two chords of ω_2 passing M and N respectively. If the segments BD, AC and MN meet at the same point K. Prove that TK bisects $\angle MTN$.

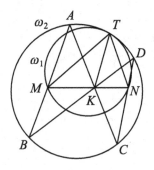

Fig. 2.1

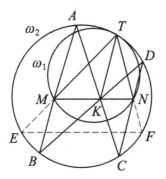

Fig. 2.2

Proof. Extend TM and TN to intersect ω_2 at E and F respectively. and join EF. It follows $MN \parallel EF$. Then

$$\frac{TM}{TN} = \frac{ME}{NF}.$$

Then by intersecting chord theorem,

$$\frac{TM^2}{TN^2} = \frac{TM}{TN} \cdot \frac{ME}{NF} = \frac{AM \cdot MB}{DN \cdot NC}. \qquad \textcircled{1}$$

In $\triangle AMK$ and $\triangle DNK$, it follows from sine rule that

$$\frac{AM}{\sin \angle AKM} = \frac{MK}{\sin \angle MAK}, \quad \frac{DN}{\sin \angle DKN} = \frac{KN}{\sin \angle KDN}.$$

Note that $\angle MAK = \angle BAC = \angle BDC = \angle KDN$, so

$$\frac{AM}{DN} = \frac{MK \cdot \sin \angle AKM}{NK \cdot \sin \angle DKN}.$$

Similarly,

$$\frac{MB}{NC} = \frac{MK \cdot \sin \angle MKB}{NK \cdot \sin \angle NKC}.$$

Thus, we have

$$\frac{AM \cdot MB}{DN \cdot NC} = \frac{MK^2}{NK^2}. \qquad \textcircled{2}$$

By $\textcircled{1}$ and $\textcircled{2}$, $\dfrac{TM^2}{TN^2} = \dfrac{MK^2}{NK^2}$, i.e. $\dfrac{TM}{TN} = \dfrac{MK}{NK}$, which implies that TK bisects $\angle MTN$. $\qquad \square$

Remark. One can also proceed with Pascal theorem, and Thales theorem.

3 Let $n \geq 2$ be an integer, and $x_1, x_2, \ldots, x_n$ are positive real numbers such that $\sum_{i=1}^{n} x_i = 1$. Prove that

$$\left(\sum_{i=1}^{n} \frac{1}{1 - x_i} \right) \left(\sum_{1 \leq i < j \leq n} x_i x_j \right) \leq \frac{n}{2}.$$

Proof (First proof). First observe that $2 \sum_{1 \leq i < j \leq n} x_i x_j = \sum_{i=1}^{n} x_i \sum_{1 \leq i < j \leq n} x_j = \sum_{i=1}^{n} x_i (1 - x_i)$, so the original inequality is equivalent to the following

$$\left(\sum_{i=1}^{n} \frac{1}{1 - x_i} \right) \left(\sum_{i=1}^{n} x_i (1 - x_i) \right) \leq n. \tag{*}$$

By rearranging the sequence, one may assume that $0 < x_1 \leq x_2 \leq \cdots \leq 1$. For any $1 \leq i < j \leq n$, we have $x_i + x_j < 1$, and $0 < x_i < x_j \leq 1$, so $(x_i - x_j)(1 - x_i - x_j) \leq 0$, i.e. $x_i(1 - x_i) \leq x_j(1 - x_j)$. It follows that $x_1(1 - x_1) \leq x_2(1 - x_2) \leq \cdots \leq x_n(1 - x_n)$. By the increasing assumption on x_i's, we have $\frac{1}{1 - x_1} \leq \frac{1}{1 - x_2} \leq \cdots \leq \frac{1}{1 - x_n}$.

By Chebyshev inequality, one has

$$\frac{1}{n} \left(\sum_{i=1}^{n} \frac{1}{1 - x_i} \right) \left(\sum_{i=1}^{n} x_i (1 - x_i) \right) \leq \left(\sum_{i=1}^{n} \left(\frac{1}{1 - x_i} \right) x_i (1 - x_i) \right) = 1,$$

and hence (*) holds.

Proof (Second proof). We first prove the following intermediate inequality: For any $1 \leq k \leq n$, we have

$$\left(2 \sum_{1 \leq i < j \leq n} x_i x_j \right) \left(\frac{1}{1 - x_k} \right) \leq 2x_k + \frac{n - 2}{n - 1} \sum_{i \neq k} x_i. \tag{**}$$

In fact, it follows from AM-GM inequality that $\sum_{i \neq k} x_i^2 \geq \frac{2}{n - 2} \sum_{i, j \neq k} x_i x_j$, and hence

$$2 \sum_{1 \leq i < j \leq n, i, j \neq k} x_i x_j \leq \frac{n - 2}{n - 1} \left(\sum_{i \neq k} x_i \right)^2$$

$$\cdot \left(2 \sum_{1 \leq i < j \leq n} x_i x_j \right) \left(\frac{1}{1 - x_k} \right)$$

$$= \left(2x_k(1 - x_k) + 2 \sum_{1 \le i < j \le n, i,j \ne k} x_i x_j \right) \left(\frac{1}{1 - x_k} \right)$$

$$= 2x_1 + \frac{2 \sum_{i \ne j \ne 1} x_i x_j}{\sum_{i \ne 1} x_i}$$

$$\le 2x_1 + \frac{n - 2}{n - 1} \sum_{i \ne 1} x_i.$$

Consequently, (**) holds. The original inequality follows if one sums up both sides of (**) with $k = 1, 2, \ldots, n$. $\qquad\square$

④ For 100 straight lines on a plane, let T be the set of all right-angled triangles bounded by some 3 lines among these 100 lines. Determine, with proof, the maximum value of $|T|$.

Proof. $|T|_{\max} = 62500$. We first prove that the number of right-angled triangles does not exceed 62500. Let $\mathcal{F}$ be the set of any 100 lines on a plane. Choose a line ℓ, let A_1 be the set of all lines parallel to ℓ, including ℓ itself. Let B_1 be the set of lines perpendicular to ℓ. In this case, B_1 could be an empty set. One can proceed inductively to define A_{i+1} and B_{i+1} by replacing $\mathcal{F}$ by $\mathcal{F} \backslash \left(\cup_{j=1}^{i} (A_j \cup B_j) \right)$. As $\mathcal{F}$ is finite, it is partitioned to mutually disjoint union of $A_1, A_2, \ldots, A_k$ and $B_1, \ldots B_k$ for some integer $k \ge 1$. Let $a_i = |A_i|$ and $b_i = |B_i| (1 \le i \le k)$. Then the number S of right-angled triangles is given by

$$S = \sum_{i=1}^{k} a_i b_i (100 - a_i - b_i)$$

$$\le \sum_{i=1}^{k} \frac{(a_j + b_j)^2}{4} \cdot (100 - a_j - b_j)$$

$$= \frac{1}{4} \sum_{i=1}^{k} (a_i + b_i) \cdot [(a_i + b_i)(100 - a_j - b_j)]$$

$$\le \frac{1}{4} \sum_{i=1}^{k} (a_i + b_i) \cdot \frac{[(a_i + b_i)(100 - a_i - b_i)]^2}{4}$$

$$= 625 \cdot \sum_{i=1}^{k} (a_i + b_i)$$

$$= 62500.$$

In the following, we give an explicit example with $S = 62500$. In the coordinate plane, let $\ell_i : x = i(1 \leq i \leq 25)$, $\ell_{i+25} : y = i(1 \leq i \leq 25)$, $\ell_{i+50} : y = x + 25 + i(1 \leq i \leq 25)$, and $\ell_{i+50} : y = -x + 100 + i(1 \leq i \leq 25)$. The set of 100 lines can be divided into 4 groups of 25 lines each, with slopes $m = 0, 1, \infty, -1$ respectively. It is easy to see that any line of the first group is perpendicular *only* to those line in the second group and vice versa. The same happens for the third and fourth group.

In this case, the number of right-angles is exactly $25 \times 25 \times 50 + 25 \times 25 \times 50 = 62500$.

In summary, the desired maximum value is 62500. □

5 Let $ABCD$ be a convex quadrilateral with area S, and $a = AB$, $b = BC$, $c = CD$, $d = DA$. For any permutation x, y, z, w of a, b, c, d, prove that

$$S_{ABCD} \leq \frac{1}{2}(xy + zw).$$

Proof. Though there are $4! = 24$ permutations on the 4 letters a, b, c and d, one can consider the following 2 cases according to x and y are the lengths of two adjacent sides of $ABCD$:

- If x and y are adjacent, then without loss of generality, it suffices to show that $S \leq \frac{1}{2}(ab + cd)$: first note that $S_{ABC} = \frac{1}{2}AB \cdot BC \sin \angle ABC \leq \frac{1}{2}ab$; and similarly, $S_{CDA} = \frac{1}{2}CD \cdot DA \sin \angle CDA \leq \frac{1}{2}cd$.
- If x and y are the lengths of two opposite sides of $ABCD$, then the inequality is $S \leq \frac{1}{2}(ac + bd)$. For this, let A' be the symmetric point of A about the perpendicular bisector of the diagonal BD. Then $\triangle BA'D \sim \triangle DAB$. Apply the previous case to $A'BCD$, we have

$$S = S_{ABCD} = S_{BAD} + S_{BCD} = S_{BA'D} + S_{BCD}$$

$$= S_{A'BCD} = S_{A'BC} + S_{CDA'}$$

$$\leq \frac{1}{2}A'B \cdot BC + \frac{1}{2}CD \cdot DA'$$

$$= \frac{1}{2}AD \cdot BC + \frac{1}{2}CD \cdot AB$$

$$= \frac{1}{2}(ac + bd).$$

One can also prove this by Ptolemy's inequality of a convex quadrilateral $ABCD$ as follows: let θ be the angle between two diagonals AC and BD, then $S_{ABCD} = \dfrac{1}{2}AC \cdot BD \sin\theta \leq \dfrac{1}{2}AC \cdot BD \leq \dfrac{1}{2}(AB \cdot CD + BC \cdot DA) = \dfrac{1}{2}(ac + bd)$.

$\square$

6 For a sequence $a_1, a_2, \ldots, a_n$ of real numbers, define the following sets

$$A = \{a_i \mid 1 \leq i \leq m\} \quad \text{and} \quad B = \{a_i + 2a_j \mid 1 \leq i, j \leq m, \ i \neq j\}.$$

Let n be a given integer and $n > 2$. For any strictly increasing arithmetic progression $a_1, a_2, \ldots, a_n$ of positive integers, determine, with proof, the minimum number of elements of set $A \triangle B$, where $A \triangle B = (A \cup B) \setminus (A \cap B)$.

Answer: Let T_n be the desired minimum value. $T_3 = 5$, and $T_n = 2n$ if $n \geq 4$.

Proof. First we prove the following

Lemma *If $n \geq 4$, and d be the common difference of the sequence $\{a_k\}$, then $B = \{3a_1 + kd \mid 1 \leq k \leq 3n - 4, \ k \in \mathbb{Z}\}$.*

Proof of lemma. In fact, for any $1 \leq i, j \leq n$, $i \neq j$, $a_i + 2a_j = 3a_1 + (i-1)d + 2(j-1)d = 3a_1 + (i + 2j - 3)d$, where $1 \leq i + 2j - 3 \leq 3n - 4$. It follows that $B \subseteq \{3a_1 + kd \mid 1 \leq k \leq 3n - 4, \ k \in \mathbb{Z}\}$.

On the other hand, for any $1 \leq k \leq 3n - 4$, we will prove that there exist i and j such that $1 \leq i, \ j \leq n$, $i \neq j$ and $k = i + 2j - 3$ as follows:

- If $k \geq 2n - 2$, choose $i = k + 3 - 2n$ and $j = n$, then $1 \leq i \leq n - 1 < j = n$, and $k = i + 2j - 3$.
- If $k \leq 2n - 3$, and k is even, choose $i = 1$, $j = \dfrac{k+2}{2}$, then $1 = i < j < n$, and $k = i + 2j - 3$.
- If $5 \leq k \leq 2n - 3$, and k is odd, choose $i = 2$, $j = \dfrac{k+1}{2}$, then $1 < i < j < n$, and $k = i + 2j - 3$.
- If $k = 1$ and 3, choose $(i, j) = (2, 1)$ and $(4, 1)$ respectively, then $1 \leq j < i \leq n$, and $k = i + 2j - 3$.

It follows that $B \supseteq \{3a_1 + kd \mid 1 \leq k \leq 3n - 4, \ k \in \mathbb{Z}\}$, and this completes the proof of the lemma.

Now we return to the proof of the original question. We first discuss the case for $n \geq 4$. Let $a_1, a_2, \ldots a_n$ be a strictly increasing arithmetic progression of positive integers with common difference $d > 0$. It is obvious $|A| = n$. It follows from lemma that $|B| = 3n - 3$.

As $a_2 = a_1 + d < 3a_1 + d$, one knows that $a_1, a_2 \notin B$, so $|A \cap B| \leq |A \backslash \{a_1, a_2\}| = n - 2$.

Then $|A \Delta B| = |A| + |B| - 2|A \cap B| \geq n + (3n - 4) - 2(n - 2) = 2n$.

On the other hand, if one sets $a_k = 1 + 2(k - 1)$, with $1 \leq k \leq n$, then $A = \{1, 3, \ldots, 2n - 3, 2n - 1\}$, and $B = \{5, 7, \ldots, 6n - 5\}$. In this case, $|A \Delta B| = 2n$, which gives the desired minimum value $T_n = 2n$ for $n \geq 4$.

Next we consider the case $n = 3$. Let $a_1 < a_2 < a_3$ be AP with common difference $d > 0$, then $|A| = 3$. Note that $2a_1 + a_2 < 2a_1 + a_3 < 2a_3 + a_1 < 2a_3 + a_2$, then $|B| \geq 4$. As $a_1, a_2 \notin B$, we have $|A \cap B| \leq 1$, so $|A \Delta B| \geq 5$. On the other hand, let $a_1 = 1$, $a_2 = 3$, $a_3 = 5$, then $A = \{1, 3, 5\}$, $B = \{5, 7, 11, 13\}$, and $|A \Delta B| = 5$. Hence, $T_3 = 5$. $\qquad \square$

7 Let $a \in (0, 1)$, $f(z) = z^2 - z + a$, $z \in \mathbb{C}$. Prove the following statement holds:

For any complex number z with $|z| \geq 1$, there exists a complex number z_0 with $|z_0| = 1$, such that $|f(z_0)| \leq |f(z)|$.

Proof. Let $U = \{z \in \mathbb{C} \mid |z| = 1\}$ and $D = \{z \in \mathbb{C} \mid |z| < 1\}$ be the set of the unit circle and the open unit disc in the complex plane $\mathbb{C}$ respectively. We first prove the following

Lemma *Let z be a complex number outside the unit circle, i.e. $|z| > 1$, then there exists a complex number $z_0 \in U$ such that $|z_0 - \omega| < |z - \omega|$ for all $\omega \in \mathbb{C}$ with $|\omega| < 1$.*

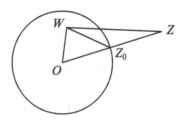

Fig. 7.1

Proof of lemma. Let $z_0 = \dfrac{z}{|z|}$, then z_0 is the intersection of U and the ray from 0 to z. As $\omega \in D$, $|\omega| < 1 = |z_0|$, and hence $\angle OZ_0W < 90°$, $\angle WZ_0Z > 90°$, so $|z_0 - \omega| < |z - \omega|$.

Returning to the original proof, we first prove that the two (complex) roots z_1, z_2 of the quadratic equation $z^2 - z + a = 0$ are in D. We discuss the following 2 cases:

- If $0 < a \leq \dfrac{1}{4}$, then the discriminant $\Delta = 1 - 4a \geq 0$, so both roots z_1, z_2 are real. It follows that z_1 and $z_2 \in (0, 1)$.
- If $\dfrac{1}{4} < a < 1$, then $\Delta = 1 - 4a < 0$, so z_1, z_2 are complex numbers and are conjugate pair. So $|z_1|^2 = |z_2|^2 = z_1 z_2 = a \in (0, 1)$.

In both cases, we know that both roots are in the open unit disc D. Then $|f(z)| = |z^2 - z - a| = |(z - z_1)(z - z_2)| = |z - z_1||z - z_2|$. Choose z_0 as follows: if $|z| = 1$, choose $z_0 = z$, then $|f(z)| = |f(z_0)|$; if $|z| > 1$, then it follows from the lemma that $z_0 = \dfrac{z}{|z|}$ satisfies $|z_0 - z_1| < |z - z_1|$, and $|z_0 - z_2| < |z - z_2|$ for all $z \in D$. Hence, $|f(z_0)|^2 = |z_0 - z_1||z_0 - z_2| < |z - z_1||z - z_2| = |f(z)|^2$, and the result follows. $\qquad\square$

8 Let k be a positive integer and $n = (2^k)!$. Prove that $\sigma(n)$ has at least a prime factor larger than 2^k, where $\sigma(n)$ is the sum of all positive divisors of n.

Proof. The highest 2-power $2^{\nu_2(n)}$ factor of n is given as

$$\nu_2(n) = \left[\frac{2^k}{2}\right] + \left[\frac{2^k}{4}\right] + \cdots + \left[\frac{2^k}{2^k}\right]$$

$$= 2^{k-1} + 2^{k-2} + \cdots + 1$$

$$= 2^k - 1.$$

Let $n = 2^{2^k - 1} p_2^{\alpha_1} p_2^{\alpha_2} \cdots p_t^{\alpha_t}$, where t is an non-negative integer, and $p_1, p_2, \ldots, p_t$ are distinct odd prime factors of n, and each of the corresponding powers $\alpha_1, \alpha_2, \ldots, \alpha_t$ is a positive integers.

Note that $\sigma(a \cdot b) = \sigma(a) \cdot \sigma(b)$ if $\gcd(a, b) = 1$. It follows from $2^{2^k - 1} \mid n$ that $\sigma(2^{2^k - 1}) \mid \sigma(n)$, and in particular, $2^{2^k} - 1 = \sigma(2^{2^k - 1}) \mid \sigma(n)$.

Let p be any prime factor of $2^{2^{k-1}} + 1$, then we check that p satisfies the condition as follows. $p \mid (2^{2^{k-1}} + 1) \mid (2^{2^k} - 1) \mid \sigma(n)$. It remains to check that $p \geq 2^k + 1$. Let d be the order of 2 modulo p, then $2^{2^{k-1}} \equiv -1 \pmod{p}$ (∗), so $2^{2^k} \equiv 1 \pmod{p}$. By the definition of order and Euclidean division, we have $d \mid 2^k$. It follows from (∗) that $d \neq 2^{k-1}$, so $d = 2^k$. It follows from Fermat's little theorem that $d \mid \phi(p) = p - 1$, and hence $2^k = d \leq p - 1$, i.e., $p \geq 2^k + 1$. $\qquad\square$

China Southeastern Mathematical Olympiad

2014 (Hangzhou, Zhejiang)

10th Grade

First Day

July 27, 2014

8:00–12:00

1. a, b, c, and d are positive integers less than an odd prime p; both $a^2 + b^2$ and $c^2 + d^2$ are multiples of p.

   Prove that exactly one of $ac + bd$ and $ad + bc$ is a multiple of p.

   (posed by Li Shenghong)

Proof. Since both $a^2 + b^2$ and $c^2 + d^2$ are multiples of p,

$$(ac + bd)(ad + bc) = (a^2 + b^2)cd + (c^2 + d^2)ab$$

is a multiple of p. Because p is a prime, so at least one of $ac+bd$ and $ad+bc$ is a multiple of p.

On the other hand, suppose that both $ac + bd$ and $ad + bc$ are multiples of p, then

$$(ac + bd) - (ad + bc) = (a - b)(c - d)$$

is a multiple of p. So $p \mid a - b$ or $p \mid c - d$.

Without loss of generality, assume $p \mid a - b$. Because $0 < a, b < p$, we have $|a - b| < p$, so $a = b$, and $a^2 + b^2 = 2a^2$ is a multiple of p. But p is an odd prime, $0 < a < p$, so $p \nmid 2a^2$. This is a contradiction. Therefore at least one of $ac + bd$ and $ad + bc$ is not a multiple of p.

In conclusion, exactly one of $ac + bd$ and $ad + bc$ is a multiple of p. $\quad\square$

2 Among $n \geq 4$ players, a ping-pong game is played between each pair of players, and no game ends with a tie.

Determine the minimum n such that, after all the games are played, no matter what the outcomes are, one can always find a tuple of 4 players (A_1, A_2, A_3, A_4) where A_i beats A_j for any $1 \leq i < j \leq 4$. (posed by He Yijie)

Proof. We first prove that, when $n = 8$, we can always find the desired 4-tuple.

There were $\binom{8}{2} = 28$ games played in total; one of the players, denoted by A_1, won $\lceil 28/8 \rceil = 4$ games. Suppose a_1, a_2, a_3, a_4 are 4 players that A_1 beat. There were 6 games played among a_1, a_2, a_3, a_4; one of them won at least $\lceil 6/4 \rceil = 2$ games among the 6. Without loss of generality, a_1 beat a_2 and a_3, and a_2 beat a_3. Set $A_2 = a_1$, $A_3 = a_2$, and $A_4 = a_3$; then the tuple (A_1, A_2, A_3, A_4) satisfies our condition.

Next, we show that when $n \leq 7$, such a tuple does not necessarily exist. It is clear that we only need to give an example when $n = 7$. Let $b_1, b_2, \ldots, b_7$ be 7 players, and we take $b_{7+k} = b_k$. Now we may have the following outcomes:

For $i = 1, 2, \ldots, 7$, b_i beat b_{i+1}, b_{i+2}, and b_{i+4}, but lost to b_{i+3}, b_{i+5}, b_{i+6}. It is easy to check that the outcome of each game is well defined. If a desired 4-tuple (A_1, A_2, A_3, A_4) existed, A_1 would be some b_i, so (A_2, A_3, A_4) is a permutation of $(b_{i+1}, b_{i+2}, b_{i+4})$. But b_{i+1} beat b_{i+2}, b_{i+2} beat b_{i+4}, and b_{i+4} beat b_{i+1}, so none of them could be A_2; a contradiction.

Thus, the minimum n is 8. $\quad\square$

3 As shown in Figure 3.1, in an abtuse triangle ABC, $AB > AC$. O is the circumcenter of the triangle; D, E, and F are the midpoints of BC, CA, and AB, respectively; the median AD intersects the lines OF and OE at M and N, respectively; the lines BM and CN intersect at P.

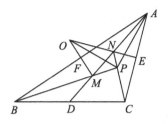

Fig. 3.1

Prove that $OP \perp AP$. (posed by Tao Pingsheng)

Proof. It is easy to see that $BM = AM$ and $CN = AN$, $\angle AMP = 2\angle BAM$, and $\angle PND = 2\angle CAM$.

As in Figure 3.2, draw OB, OC. Since O is the circumcenter of $\triangle ABC$, $\angle BOC = 2\angle BAC = 2\angle BAM + 2\angle CAM = \angle AMP + \angle PND = \angle BPC$.

So, B, O, P, and C are concyclic, and
$$\angle BPO = \angle BCO = \angle CBO = \angle OPN.$$
On the other hand, apply Menelaus' theorem to $\triangle BCP$ with the transversal line DMN, we get
$$\frac{BD}{DC} \cdot \frac{CN}{NP} \cdot \frac{PM}{MB} = 1.$$
This, combined with the facts $BD = DC$, $BM = AM$, and $CN = AN$, gives
$$\frac{PM}{PN} = \frac{MB}{NC} = \frac{AM}{AN}.$$
So, PA is the exterior bisector of $\angle MPN$. We proved that PO is the bisector of $\angle MPN$. Therefore $OP \perp AP$.

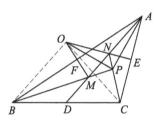

Fig. 3.2 □

4 n is a positive integer; non-negative reals $x_1, x_2, \ldots, x_n$ satisfy $x_i x_j \leq 4^{-|i-j|}$ for any $1 \leq i, j \leq n$.
Prove that

$$x_1 + x_2 + \ldots x_n < \frac{5}{3}.$$

(posed by Jin Mengwei)

Proof (First proof). Let $M = \max_{1 \leq i \leq n} x_i$.

Case 1. $0 \leq M \leq \frac{2}{3}$. We prove that, for any s, k, where $1 \leq s \leq s + k \leq n$, we have

$$\sum_{i=s}^{s+k} x_i \leq \frac{2}{3} + \sum_{i=1}^{k} \frac{1}{2^i}; \qquad \text{①}$$

here we agree on

$$\sum_{i=1}^{0} \frac{1}{2^i} = 0.$$

We prove this by induction on k. When $k = 0$, $x_s \leq M \leq \frac{2}{3}$ for any $1 \leq s \leq n$.

Suppose ① holds for a particular k and any such s. For any s, k where $1 \leq s \leq s + k + 1 \leq n$, we have

$$\min\{x_s, x_{k+1+s}\} \leq \sqrt{x_s x_{k+1+s}} \leq \sqrt{4^{-|s-(k+1+s)|}} = \frac{1}{2^{k+1}}.$$

When $x_{k+1+s} \leq \frac{1}{2^{k+1}}$, by inductive hypothesis,

$$\sum_{i=s}^{s+k+1} x_i = \left(\sum_{i=s}^{s+k} x_i \right) + x_{s+k+1}$$

$$\leq \left(\frac{2}{3} + \sum_{i=1}^{k} \frac{1}{2^i} \right) + \frac{1}{2^{k+1}}$$

$$= \frac{2}{3} + \sum_{i=1}^{k+1} \frac{1}{2^i};$$

similarly, when $x_k \leq \dfrac{1}{2^{k+1}}$, by the inductive hypothesis,

$$\sum_{i=s}^{s+k+1} x_i = x_i + \left(\sum_{i=s+1}^{s+k+1} x_i\right)$$

$$\leq \frac{1}{2^{k+1}} + \left(\frac{2}{3} + \sum_{i=1}^{k} \frac{1}{2^i}\right)$$

$$= \frac{2}{3} + \sum_{i=1}^{k+1} \frac{1}{2^i}.$$

So ① holds when k is replaced by $k+1$; and this completes the proof of ① . In particular,

$$x_1 + x_2 + \cdots + x_n \leq \frac{2}{3} + \sum_{i=1}^{n-1} \frac{1}{2^i} < \frac{2}{3} + 1 = \frac{5}{3}.$$

Case 2. $M \geq \dfrac{2}{3}$. Suppose $x_t = M$, we have $M = \sqrt{x_t x_t} \leq \sqrt{4^{-|t-t|}} = 1$, and, for any $1 \leq i \leq n$

$$x_i \leq \frac{4^{-|i-t|}}{x_t} = \frac{1}{4^{|i-t|}M}.$$

So

$$x_1 + x_2 + \cdots + x_n = x_t + \sum_{i=1}^{t-1} x_i + \sum_{i=t+1}^{n} x_i$$

$$\leq M + \sum_{i=1}^{t-1} \frac{1}{4^{t-i}M} + \sum_{i=t+1}^{n} \frac{1}{4^{i-t}M} \qquad ②$$

$$< M + \frac{2}{M} \sum_{j=1}^{\infty} \frac{1}{4^j} = M + \frac{2}{3M}.$$

Note that, when $\dfrac{2}{3} < M \leq 1$,

$$\left(M + \frac{2}{3M}\right) - \frac{5}{3} = \frac{(M-1)(3M-2)}{3M} \leq 0.$$

So, by ② ,

$$x_1 + x_2 + \cdots + x_n < \frac{5}{3}.$$

Proof (Second proof, based on the solutions from the contestants).
When $i = j$, the statements implies $x_i^2 \leq 1$, so

$$0 \leq x_i \leq 1, \quad (1 \leq i \leq n).$$

Let $S = \sum_{i=1}^{n} x_i$, and, for any $0 \leq j \leq n$, $S_j = \sum_{i=1}^{j} x_i$, so

$$0 = S_0 \leq S_1 \leq \cdots \leq S_n = S.$$

Therefore, there exists some k, $0 \leq k \leq n - 1$, where $S_k \leq S/2 \leq S_{k+1}$.
Let $T_k = S - S_k$, $T_{k+1} = S - S_{k+1}$, then

$$|S_k - T_k| + |S_{k+1} - T_{k+1}| = |2S_k - S| + |2S_{k+1} - S|$$
$$= S - 2S_k + 2S_{k+1} - S = 2x_{k+1} \leq 2.$$

So there exists $l \in \{k, k+1\}$, such that

$$|S_l - T_l| \leq 1. \qquad (3)$$

On the other hand, we have

$$S_l T_l = \sum_{i=1}^{l} x_i \cdot \sum_{j=l+1}^{n} x_j = \sum_{i=1}^{l} \sum_{j=l+1}^{n} x_i x_j$$

$$\leq \sum_{i=1}^{l} \sum_{j=l+1}^{n} 4^{-(j-i)}$$

$$= \sum_{i=1}^{l} \frac{1}{4^{l-i}} \sum_{j=l+1}^{n} \frac{1}{4^{j-l}} \qquad (4)$$

$$< \sum_{r=0}^{\infty} \frac{1}{4^r} \sum_{r=1}^{\infty} \frac{1}{4^r}$$

$$= \frac{4}{3} \times \frac{1}{3} = \frac{4}{9}.$$

From (3) and (4), we get

$$\sum_{i=1}^{n} x_i = S_l + T_l = \sqrt{(S_l - T_l)^2 + 4S_l T_l} < \sqrt{1 + 4 \times \frac{4}{9}} = \frac{5}{3}.$$

Second Day

July 28, 2014

8:00–12:00

5 $\triangle ABC$ and $\triangle XYZ$ are two acute triangles. Prove that the maximum number among $\cot A(\cot Y + \cot Z)$, $\cot B(\cot Z + \cot X)$, and $\cot C(\cot X + \cot Y)$ is no less than $2/3$. (posed by Zhang Sihui)

Proof. Denote $\cot A, \cot B, \cot C, \cot X, \cot Y$, and $\cot Z$ by a, b, c, x, y, and z, respectively. We have

$$ab + bc + ca = xy + yz + zx = 1,$$
$$a, b, c, x, y, z > 0.$$

By Cauchy-Schwarz,

$$(a + b + c)^2(x + y + z)^2 = (a^2 + b^2 + c^2 + 2)(x^2 + y^2 + z^2 + 2)$$
$$\geq (ax + by + cz + 2)^2,$$

i.e.,

$$(a + b + c)(x + y + z) \geq ab + by + cz + 2,$$

that is,

$$a(y + z) + b(z + x) + c(x + y) \geq 2.$$

Hence,

$$\max\{a(y + z), b(z + x), c(x + y)\} \geq \frac{2}{3}. \qquad \square$$

6 Integers a, b, and c, and a real number r satisfy $ar^2 + br + c = 0$ and $ac \neq 0$. Prove that $\sqrt{r^2 + c^2}$ is irrational. (posed by He Yijie)

Proof. From the condition we know that $b^2 - 4ac \geq 0$.
Let $r = \dfrac{-b + m}{2a}$, where $m = \sqrt{b^2 - 4ac}$. Since $ac \neq 0$, we have

$$m \neq \pm b. \qquad \text{①}$$

Assume, for the sake of a contradiction, $\sqrt{r^2 + c^2}$ is a rational q, and let $s = 2aq \in \mathbb{Q}$, then

$$s^2 = 4a^2q^2 = 4a^2(r^2 + c^2)$$
$$= (2ar)^2 + 4a^2c^2 \qquad \textcircled{2}$$
$$= (m - b)^2 + 4a^2c^2 > 0.$$

If $m \in \mathbb{Z}$, the right hand side of the above is an integer, so $s \in \mathbb{Z}$. However,

$$4s^2 = 4(m - b)^2 + (4ac)^2$$
$$= 4(m - b)^2 + (b^2 - m^2)^2$$
$$= (m - b)^2[4 + (m + b)^2],$$

so $4 + (m + b)^2$ is a square, which implies $m + b = 0$, contradicts $\textcircled{1}$.

So $m \notin \mathbb{Z}$. Note that $m^2 = b^2 - 4ac \in \mathbb{Z}$, so $m \notin \mathbb{Q}$. And, by $\textcircled{2}$,

$$2mb = m^2 + b^2 + 4a^2c^2 - s^2 \in \mathbb{Q},$$

therefore $b = 0$; then

$$s^2 + 1 = m^2 + 4a^2c^2 + 1 = -4ac + 4a^2c^2 + 1 = (2ac - 1)^2.$$

So $s^2 + 1$ is a square, which implies $s = 0$, contradicts $\textcircled{2}$.

In conclusion, the assumption that $\sqrt{r^2 + c^2}$ is rational does not hold; it must be irrational. $\qquad\qquad\square$

7 As shown in Figure 7.1, O is the center of a fixed circle ω_1 in the plane. P is a point on the circle. A circle ω_2 has P as its center, and radius less than the radius of ω_1. ω_2 intersects ω_1 at T and Q. TR

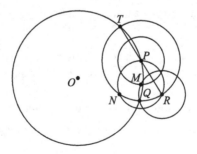

Fig. 7.1

is a diameter of ω_2. Two circles with radius equal to RQ and centers R and P, respectively, intersect at a point M which is on the same side of PR as Q. The circle with center M and radius MR intersects ω_2 at R and N.

Prove that the circle with center T and radius TN passes through O.

(posed by Zhang Sihui)

Proof. As in Figure 7.2, let $\angle QTR = \alpha$. By the assumptions, $\overparen{TP} = \overparen{PQ}$, so $\angle TOP = 2\angle TQP = 2\angle QTR = 2\alpha$. Let $QR = x$ and $TP = r$, then

$$\frac{x}{2r} = \frac{QR}{TR} = \sin\alpha = \frac{TP/2}{OT} = \frac{r/2}{OT},$$

therefore $OT = \dfrac{r^2}{x}$.

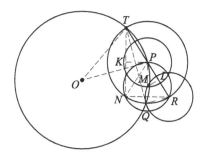

Fig. 7.2

Draw $PK \perp TN$ with the intersection point K, $ML \perp TR$ with the intersection point L; then

$$\angle TPK = \angle KPN. \qquad \textcircled{1}$$

Note that $NM = PM = RM = QR = x$, so $\triangle PNM \cong \triangle PRM$, therefore

$$\angle NPM = \angle RPM. \qquad \textcircled{2}$$

From $\textcircled{1}$ and $\textcircled{2}$, $\angle KPN + \angle NPM = 90°$; yet

$$\angle PML + \angle NPM = \angle PML + \angle RPM = 90°,$$

so $\angle KPN = \angle PML$. Therefore

$$\frac{TN/2}{r} = \sin\angle KPN = \sin\angle PML = \frac{r/2}{x}.$$

Thus we have $TN = \dfrac{r^2}{x} = OT$; the circle with center T and radius TN passes through O. □

8 As shown in Figure 8.1, a unit square P has a squares on its left, b squares on its right, c squares above it and d squares below it. When $(a - b)(c - d) = 0$, we call such a configuration of squares a *cross star*. 2014 unit squares form a rectangular grid of dimensions 38×53. Determine the number of cross stars one can find in such a grid.

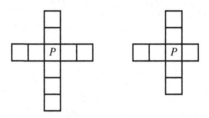

Fig. 8.1

(posed by Tao Pingsheng)

Solution For a cross star, call the square P in the statement its *center*; call it *standing* if $a = b$, call it *lying* if $c = d$. Some cross stars can be both standing and lying.

For a rectangle R consists of unit squares on the grid and a cross star S, we call R *the meta-rectangle* of S if S is the union of one row and one column of R. In a meta-rectangle, a cross star S is uniquely determined by its center. Cross stars corresponding to different meta-rectangles are different. By definition, if R is the meta-rectangle for any cross star, it has at least 3 rows and 3 columns; any corresponding cross star is not lying if R has an even number of rows, and not standing if R has an even number of columns.

Now consider the grid $P_{m,n}$ with m rows and n columns of unit squares. Denote by A, B, and C the number of cross stars in $P_{m,n}$ that are lying, standing, and both lying and standing, respectively.

In $P_{m,n}$, for k and l such that $3 \le 2k + 1 \le m$ and $3 \le l \le n$, we have $(m - 2k)(n - l + 1)$ rectangles with $2k + 1$ rows and l columns. For every such rectangle R, the center of a corresponding lying cross star must be on the $(k + 1)$-st row and between 2nd and $(l - 1)$-th column, inclusive. So there are $l - 2$ lying corss stars having R as the meta-rectangle, and

$$A = \sum_{k=1}^{\lfloor \frac{m-1}{2} \rfloor} \sum_{l=3}^{n} (l - 2)(m - 2k)(n - l + 1)$$

$$= \sum_{k=1}^{\lfloor \frac{m-1}{2} \rfloor} (m - 2k) \cdot \sum_{l=1}^{n-2} l(n - l - 1).$$

We calculate

$$\sum_{k=1}^{\lfloor \frac{m-1}{2} \rfloor} (m - 2k) = \left\lfloor \frac{m-1}{2} \right\rfloor \cdot \frac{(m - 2) + (m - 2 \lfloor \frac{m-1}{2} \rfloor)}{2}$$

$$= \left\lfloor \frac{m-1}{2} \right\rfloor \left(m - \left\lfloor \frac{m+1}{2} \right\rfloor \right),$$

and

$$\sum_{l=1}^{n-2} l(n - l - 1) = \left(\sum_{l=1}^{n-2} l \right)(n - 1) - \sum_{l=1}^{n-2} l^2$$

$$= \frac{(n - 2)(n - 1)^2}{2} - \frac{(n - 2)(n - 1)(2n - 3)}{6}$$

$$= \frac{n(n - 1)(n - 2)}{6}.$$

So

$$A = \left\lfloor \frac{m-1}{2} \right\rfloor \left(m - \left\lfloor \frac{m+1}{2} \right\rfloor \right) \frac{n(n - 1)(n - 2)}{6}. \qquad ①$$

Similarly, by symmetry, interchanging the rows and columns in the analysis above, we get

$$B = \left\lfloor \frac{n-1}{2} \right\rfloor \left(n - \left\lfloor \frac{n+1}{2} \right\rfloor \right) \frac{m(m - 1)(m - 2)}{6}. \qquad ②$$

Each rectangle with $2k+1$ rows and $2l+1$ columns is the meta-rectangle of exactly one cross star that is both standing and lying, where the star's center is the central square of the rectangle. There are $(m - 2k)(n - 2l)$

such rectangles, so

$$C = \sum_{k=1}^{\lfloor \frac{m-1}{2} \rfloor} \sum_{l=1}^{\lfloor \frac{n-1}{2} \rfloor} (m - 2k)(n - 2l)$$

$$= \left\lfloor \frac{m-1}{2} \right\rfloor \left(m - \left\lfloor \frac{m+1}{2} \right\rfloor \right) \left\lfloor \frac{n-1}{2} \right\rfloor \left(n - \left\lfloor \frac{n+1}{2} \right\rfloor \right).$$

(3)

In particular, when $m = 38$ and $n = 53$, we have

$$\left\lfloor \frac{m-1}{2} \right\rfloor \left(m - \left\lfloor \frac{m+1}{2} \right\rfloor \right) = 18 \times 19 = 342,$$

$$\frac{m(m-1)(m-2)}{6} = 8,436,$$

$$\left\lfloor \frac{n-1}{2} \right\rfloor \left(n - \left\lfloor \frac{n+1}{2} \right\rfloor \right) = 26 \times 26 = 676,$$

$$\frac{n(n-1)(n-2)}{6} = 23,426.$$

Plug in (1), (2), and (3), and we get

$$A = 342 \times 23,426 = 8,011,692,$$

$$B = 676 \times 8,436 = 5,702,736,$$

and

$$C = 342 \times 676 = 231,192.$$

Thus, by the inclusion-exclusion principle, the number of cross stars in $P_{38,53}$ is $A + B - C = 13,483,236.$ □

11th Grade

First Day

July 27, 2014

8:00–12:00

1. As in Figure 1.1, in an acute triangle ABC, $AB > AC$, M is the midpoint of BC, I is the incenter, MI intersects AC at D, and BI meets the circumcircle at another point E.

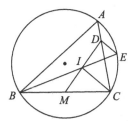

Fig. 1.1

Prove that $\dfrac{ED}{EI} = \dfrac{IC}{IB}$. (posed by Zhang Peng)

Proof. Let $BC = a$, $CA = b$, and $AB = c$. As in Figure 1.2, let F be the intersection of BE and AC, and draw AI and CE.

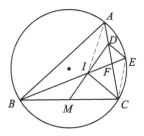

Fig. 1.2

Apply Menelaus' theorem to $\triangle BCF$ with the transversal line MID, we get $\dfrac{BM}{MC} \cdot \dfrac{CD}{DF} \cdot \dfrac{FI}{IB} = 1$; since $BM = MC$, so $\dfrac{CD}{DF} = \dfrac{IB}{FI}$.

Since AI bisects $\angle BAC$, $\dfrac{IB}{FI} = \dfrac{AB}{AF}$.

Since A, B, C, and E are concyclic, $\angle ABE = \angle ACE$, so $\triangle ABF \sim \triangle ECF$, therefore $\dfrac{AB}{AF} = \dfrac{EC}{EF}$.

Noting that

$$\angle EBC = \angle EBA = \angle ECA, \qquad \qquad \text{④}$$

so

$$\angle EIC = \angle EBC + \angle BCI = \angle ECA + \angle ICA = \angle ECI,$$

therefore

$$EC = EI.$$

We now have

$$\frac{CD}{DF} = \frac{IB}{FI} = \frac{AB}{AF} = \frac{EC}{EF} = \frac{EI}{EF},$$

so $ED \parallel IC$, which implies

$$\angle BCI = \angle ICD = \angle CDE.$$

Together with ④ we get $\triangle BCI \sim \triangle CDE$, so $\dfrac{IC}{IB} = \dfrac{ED}{EC} = \dfrac{ED}{EI}$. $\qquad \square$

2 Among $n \geq 4$ players, a ping-pong game is played between each pair of players, and no game ends with a tie.

Determine the minimum n such that, after all the games are played, no matter what the outcomes are, one can always find a tuple of 4 players (A_1, A_2, A_3, A_4) where A_i beats A_j for any $1 \leq i < j \leq 4$. (posed by He Yijie)

The problem is identical to Problem 2, 10th grade; so is the solution.

3 p is a prime number, and x, y, and z are positive integers satisfying $x < y < z < p$ and

$$\left\{\frac{x^3}{p}\right\} = \left\{\frac{y^3}{p}\right\} = \left\{\frac{z^3}{p}\right\},$$

where $\{a\}$ denote the fractional part of a real a. Prove that

$$(x + y + z) \mid (x^5 + y^5 + z^5).$$

(posed by Yang Xiaoming)

Proof. Obviously $p > 3$, and none of $x - y$, $y - z$, and $z - x$ is a multiple of p.

Since
$$\left\{\frac{x^3}{p}\right\} = \left\{\frac{y^3}{p}\right\} = \left\{\frac{z^3}{p}\right\},$$
we have $p \mid x^3 - y^3 = (x - y)(x^2 + xy + y^2)$, so $p \mid (x^2 + xy + y^2)$. Similarly, $p \mid (y^2 + yz + z^2)$ and $p \mid (z^2 + zx + x^2)$.

So
$$(x^2 + xy + y^2) - (y^2 + yz + z^2)$$
$$= x^2 - z^2 + xy - yz$$
$$= (x - z)(x + y + z) \equiv 0 \pmod{p}.$$

Let $A = x + y + z$, $B = x^2 + y^2 + z^2$, and $C = xy + yz + zx$, then $p \mid A$ and
$$B + 2C = (x + y + z)^2 \equiv 0 \pmod{p},$$

$$2B + C$$
$$= (x^2 + xy + y^2) + (y^2 + yz + z^2) + (z^2 + zx + x^2)$$
$$\equiv 0 \pmod{p}.$$

So
$$p \mid 2(2B + C) - (B + 2C) = 3B.$$

Since $(p, 3) = 1$, we have $p \mid B$ and then $p \mid C$. Hence,
$$x^5 + y^5 + z^5$$
$$= (x^3 + y^3 + z^3)(x^2 + y^2 + z^2) - x^2 y^2 (x + y)$$
$$\quad - y^2 z^2 (y + z) - z^2 x^2 (z + x) \qquad \text{(5)}$$
$$= (x^3 + y^3 + z^3)B - x^2 y^2 (A - z) - y^2 z^2 (A - x) - z^2 x^2 (A - y)$$
$$\equiv x^2 y^2 z + y^2 z^2 x + z^2 x^2 y = Cxyz \equiv 0 \pmod{p}.$$

Because $x + y + z < 3p$, so $x + y + z = p$ or else $x + y + z = 2p$.

When $x + y + z = p$, we get $(x + y + z) \mid (x^5 + y^5 + z^5)$.

When $x + y + z = 2p$,
$$x^5 + y^5 + z^5 \equiv x + y + z \equiv 0 \pmod{2},$$
by (5) and the fact that p is odd, we also have $(x + y + z) \mid (x^5 + y^5 + z^5)$. $\square$

4   n is a positive integer; non-negative reals $x_1, x_2, \ldots, x_n$ satisfy $x_i x_j \leq 4^{-|i-j|}$ for any $1 \leq i, j \leq n$.

Prove that

$$x_1 + x_2 + \ldots x_n < \frac{5}{3}.$$

(posed by Jin Mengwei)

The problem is identical to Problem 4, 10th grade; so is the solution.

Second Day

July 28, 2014

8:00–12:00

5 n is an integer bigger than 1; real numbers $x_1, x_2, \ldots, x_n$ satisfy $x_1 + x_2 + \cdots + x_n = 1$. Prove that

$$\sum_{i=1}^{n} \frac{x_i}{x_{i+1} - x_{i+1}^3} \geq \frac{n^3}{n^2 - 1},$$

here $x_{n+1} = x_1$. (posed by Li Shenghong)

Proof. Obviously $0 < x_i < 1$ for each $i = 1, 2, \ldots, n$. By Cauchy-Schwarz and the AM-GM inequalities,

$$\left(\sum_{i=1}^{n} \frac{x_i}{x_{i+1} - x_{i+1}^3} \right) \cdot \left(\sum_{i=1}^{n} (1 - x_{i+1}^2) \right)$$

$$\geq \left(\sum_{i=1}^{n} \sqrt{\frac{x_i}{x_{i+1} - x_{i+1}^3}} \cdot \sqrt{1 - x_{i+1}^2} \right)^2$$

$$= \left(\sum_{i=1}^{n} \sqrt{\frac{x_i}{x_{i+1}}} \right)^2 \qquad (6)$$

$$\geq \left(n \cdot \left(\prod_{i=1}^{n} \sqrt{\frac{x_i}{x_{i+1}}} \right)^{\frac{1}{n}} \right)^2 = n^2,$$

and

$$\sum_{i=1}^{n} (1 - x_{i+1}^2) = n - \sum_{i=1}^{n} x_i^2 \leq n - \frac{1}{n} \cdot \left(\sum_{i=1}^{n} x_i \right)^2 = \frac{n^2 - 1}{n}. \qquad (7)$$

By (6) and (7),

$$\sum_{i=1}^{n} \frac{x_i}{x_{i+1} - x_{i+1}^3} \geq \frac{n^3}{n^2 - 1}, \qquad \square$$

6 As shown in Figure 6.1, O is the center of a fixed circle ω_1 in the plane. P is a point on the circle. A circle ω_2 has P as its center, and radius less than the radius of ω_1. ω_2 intersects ω_1 at T and Q. TR

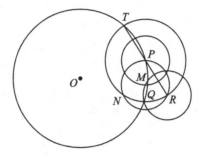

Fig. 6.1

is a diameter of ω_2. Two circles with radius equal to RQ and centers R and P, respectively, intersect at a point M which is on the same side of PR as Q. The circle with center M and radius MR intersects ω_2 at R and N.

Prove that the circle with center T and radius TN passes through O. (posed by Zhang Sihui)

The problem is identical to Problem 7, 10th grade; so is the solution.

7 Prove that the equation $a^2 + b^3 = c^4$ has an infinite sequence of distinct solutions in positive integers (a_i, b_i, c_i), $i = 1, 2, \ldots$, such that c_n and c_{n+1} are coprime for any positive integer n. (posed by Tao Pingsheng)

Proof. The equation in the statement is equivalent to $b^3 = (c^2 - a)(c^2 + a)$. We consider the solutions (a, b, c) where $b = c^2 - a$, $b^2 = c^2 + a$, and b is odd. For these solutions,

$$2c^2 = b(b+1), \quad 2a = b(b-1).$$

b is odd, so

$$c^2 = b \cdot \frac{b+1}{2}, \quad b, \; \frac{b+1}{2} \in \mathbb{N}.$$

We can take

$$b = x^2, \quad \frac{b+1}{2} = y^2, \quad c = xy,$$

and

$$a = \frac{b(b-1)}{2} = x^2(y^2 - 1),$$

as long as positive integers x and y satisfy

$$x^2 - 2y^2 = -1. \tag{8}$$

When $y \geq 2$, the corresponding (a, b, c) is a positive solution triple to our equation.

Start from two solutions of (8),

$$(x_1, y_1) = (7, 5), \quad (x_2, y_2) = (41, 29)$$

we get two solutions to our equation

$$(a_1, b_1, c_1) = (1176, 49, 35),$$

$$(a_2, b_2, c_2) = (1412040, 1681, 1189).$$

Here $(c_1, c_2) = 1$.

Multiply both sides of $a^2 + b^3 = c^4$ by k^{12}, we get

$$(ak^6)^2 + (bk^4)^3 = (ck^3)^4,$$

so (ak^6, bk^4, ck^3) is a solution to our equation whenever (a, b, c) is.

Hence, we pick primes

$$41 < p_1 < p_2 < p_3 < \ldots,$$

and, for each $j = 1, 2, \ldots$, define

$$(a_{2j+1}, b_{2j+1}, c_{2j+1}) = (a_1 p_{2j-1}^6, b_1 p_{2j-1}^4, c_1 p_{2j-1}^3),$$

$$(a_{2j+2}, b_{2j+2}, c_{2j+2}) = (a_2 p_{2j}^6, b_2 p_{2j}^4, c_2 p_{2j}^3).$$

Since the p_i's are distinct primes and are all coprime to $c_1 = 7 \times 5$ and $c_2 = 41 \times 29$, so

$$(c_{2j}, c_{2j+1}) = (c_{2j+1}, c_{2j+2}) = (c_1, c_2) = 1.$$

Thus the sequence of solutions (a_1, b_i, c_i) we defined has the desired property. □

8 As shown in Figure 8.1, a unit square P has a squares on its left, b squares on its right, c squares above it and d squares below it. When $(a - b)(c - d) = 0$, we call such a configuration of squares a *cross star*. 2014 unit squares form a rectangular grid of dimensions 38×53. Determine the number of cross stars one can find in such a grid.

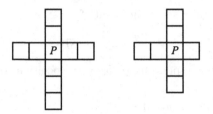

Fig. 8.1

(posed by Tao Pingsheng)

The problem is identical to Problem 8, 10th grade; so is the solution.

China Southeastern Mathematical Olympiad

2015 (Fuan, Fujian)

10th Grade

First Day

July 27, 2015

8:00–12:00

1. A series $\{a_n\}$ satisfies $a_1 = 1$, $a_{2k} = a_{2k-1} + a_k$, and $a_{2k+1} = a_{2k}$ for all $k = 1, 2, \ldots$.
Prove that

$$a_{2^n} < 2^{\frac{n^2}{2}}$$

for any integer $n \geq 3$.

Proof. We have $a_4 = a_3 + a_2 = 2a_2 = 4a_1 = 4$.
For any integer $i \geq 2$,

$$a_{2i} - a_{2i-2} = (a_{2i-1} + a_i) - a_{2i-1} = a_i,$$

and since $\{a_n\}$ is monotone non-decreasing, so for any positive integer m, we have

$$a_{2^{m+1}} - a_{2^m} = \sum_{i=2^{m-1}+1}^{2^m} (a_{2i} - a_{2i-2}) = \sum_{i=2^{m-1}+1}^{2^m} a_i \leq 2^{m-1} a_{2^m}.$$

151

So $\dfrac{a_{2m+1}}{a_{2m}} \leq 1 + 2^{m-1} \leq 2^m$. Therefore, for any integer $n \geq 3$,

$$a_{2^n} = \left(\prod_{m=2}^{n-1} \frac{a_{2m+1}}{a_{2m}} \right) \times a_4 \leq \left(\prod_{m=2}^{n-1} 2^m \right) \times 4$$

$$= 2^{(n-1)+(n-2)+\cdots+2+2} = 2^{\frac{n^2-n+2}{2}} < 2^{\frac{n^2}{2}}. \qquad \square$$

2 In $\triangle ABC$, $AB > AC$; I is its incenter. The circle Γ has AI as its diameter and intersects the circumcircle of $\triangle ABC$ at A and D, where D is on the arc $\overset{\frown}{AC}$ that does not contain B. The line passing through A and parallel to BC intersects Γ at A and another point E, as in Figure 2.1. Suppose DI bisects $\angle CDE$, and $\angle ABC = 33°$. Determine the degree of $\angle BAC$.

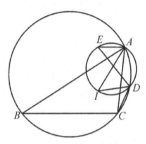

Fig. 2.1

Solution Denote the circumcircle of $\triangle ABC$ by Γ_1. Extend AI and meet Γ_1 at M; extend DE and meet Γ_1 at N; draw the segments AD, EI, and MN, as in Figure 2.2.

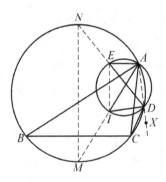

Fig. 2.2

Since A, D, I, and E are concyclic, and A, D, M, and N are concyclic, we have $\angle AIE = \angle ADE = \angle ADN = \angle AMN$, so $EI \parallel MN$.

AI is the diameter of Γ, so $EI \perp AE$, yet $AE \parallel BC$, so $EI \perp BC$, and then $MN \perp BC$.

Since I is the incenter of $\triangle ABC$, M is the midpoint of the arc $\overset{\frown}{BC}$ that does not contain A, so N is the midpoint of $\overset{\frown}{BAC}$, and

$$\overset{\frown}{NB} = \overset{\frown}{NC}.$$

Let X be any point on the extension of the segment AD.

By the assumption, $ID \perp AX$, together with the fact that DI bisects $\angle CDE$, we get

$$\angle ADE = \angle CDX = \angle ABC,$$

therefore

$$\overset{\frown}{NA} = \overset{\frown}{AC}.$$

Hence

$$\overset{\frown}{ANB} = \overset{\frown}{NB} + \overset{\frown}{NA} = \overset{\frown}{NC} + \overset{\frown}{NA} = 3\overset{\frown}{AC},$$

therefore

$$\angle ACB = 3\angle ABC.$$

Since $\angle ABC = 33°$, we have

$$\angle BAC = 180° - \angle ABC - \angle ACB = 180° - 4 \times 33° = 48°. \qquad \square$$

3 Is it possible to arrange the first 2015 positive integers on a circle so that the sum of any two adjacent numbers is either divisible by 4 or divisible by 7? Justify your answer.

Solution   We prove that the answer is positive.

Define the three sequences

$$A = (1, 3, 5, 7, \ldots, 2013, 2015),$$

$$B = (8, 4, 2012, 2008, \ldots, 16, 12),$$

and

$$C = (2, 2014, 2010, 2006, \ldots, 10, 6).$$

Any adjacent pair in each of the sequences has sum that is a multiple of 4.

Now arrange A, B, and C, in this order, on the circle, we get an arrangement of the first 2015 positive integers. 2015, the last element of A, is followed by 8, the first element of B; 12, the last element of B, is followed by 2, the first element of C; and 6, the last element of C, is followed by 1, the first element of A. Note that $2015 + 8$, $12 + 2$, $6 + 1$ are all multiples of 7.

So any adjacent pair on the circle has a sum that is a multiple of 4 or 7.

$\square$

4 For each positive integer n, define $P_n = \{n^k : k = 0, 1, 2, \ldots\}$. We call a triple (a, b, c) of positive integers *lucky* if there is a positive integer m such that the three (not necessarily distinct) numbers $a - 1$, $ab - 12$, and $abc - 2015$ all belong to P_m.

Determine the number of lucky triples.

Solution We first discuss necessary conditions for a triple (a, b, c) to be lucky.

Suppose a positive integer m and three non-negative integers α, β, and γ satisfy

$$a - 1 = m^\alpha, \tag{1}$$

$$ab - 12 = m^\beta, \tag{2}$$

and

$$abc - 2015 = m^\gamma. \tag{3}$$

We prove the following.

(a) m is even.

If m is odd, by (1) we have a is even; then the left hand side of (2) is even, but the right hand side is odd, a contradiction.

(b) $\gamma = 0$.

Assume for the sake of a contradiction $\gamma > 0$. By (3), $abc = 2015 + m^\gamma$ is odd, so ab is odd, and, by (2), $m^\beta = ab - 12$ is odd. Yet m is even, so we must have $ab - 12 = 1$ and $ab = 13$.

From (1), $a > 1$, so $a = 13$, and $m^\alpha = a - 1 = 12$, therefore $m = 12$. Now, by (3), $12^\gamma = abc - 2015 = 13(c - 155)$, which is impossible.

So $\gamma = 0$, and

$$abc = 2016. \tag{4}$$

(c) $\alpha = 0$.

Assume $\alpha > 0$. By ①, a is odd and bigger than 1; and by ④, a is a divisor of 2016. Note that $2016 = 2^5 \times 3^2 \times 7$, so a is one of $3, 7, 9, 21, 63$.

When a is one of $3, 9, 21$, and 63, we have $3 \mid a$ so $3 \mid ab - 12$; by ②, $3 \mid m$. However, by ①, $m^\alpha = a - 1 = 2 \pmod 3$, a contradiction.

When $a = 7$, by ①, $m^\alpha = a - 1 = 6$, so $m = 6$. Now ② becomes $7b - 12 = 6^\beta$; however, $6^\beta \equiv \pm 1 \pmod 7$, a contradiction.

So $\alpha = 0$, and $a = 2$; together with ④ we get $bc = 1008$. Now ② becomes

$$2b - 12 = m^\beta. \qquad\qquad ⑤$$

So $b > 6$. Conversely, when $b > 6$, we can take $m = 2b - 12$ and $\beta = 1$.

So, (a, b, c) is a lucky triple if and only if

$$a = 2, bc = 1008, b > 6.$$

$1008 = 2^4 \times 3^2 \times 7$ has $(4+1) \times (2+1) \times (1+1) = 30$ positive divisors, five of which — 1, 2, 3, 4, 6 — are not greater than 6, so there are $30 - 5 = 25$ possible values for b. Thus, the number of lucky triples is 25. $\qquad \square$

Second Day

July 28, 2015

8:00–12:00

5 a and b are real numbers; and the function $f(x) = ax + b$ satisfies that $|f(x)| \leq 1$ for any $x \in [0, 1]$.
Determine the range of $S = (a+1)(b+1)$.

Solution   Let $t = a + b$. We have $b = f(0) \in [-1, 1]$, $t = f(1) \in [-1, 1]$, and

$$S = (a+1)(b+1) = (t - b + 1)(b + 1).$$

View the right hand side of the above equation as a linear function

$$g(t) = (b+1)t + (1 - b^2), \quad t \in [-1, 1]$$

in t. Since $b + 1 \geq 0$, $g(-1) \leq g(t) \leq g(1)$, i.e.,

$$-b^2 - b \leq g(t) \leq -b^2 + b + 2. \tag{1}$$

When $b \in [-1, 1]$, we have

$$-b^2 - b = -\left(b + \frac{1}{2}\right)^2 + \frac{1}{4} \geq -\left(1 + \frac{1}{2}\right)^2 + \frac{1}{4} = -2,$$

and

$$-b^2 + b + 2 = -\left(b - \frac{1}{2}\right)^2 + \frac{9}{4} \leq \frac{9}{4}.$$

Together with ①, $S = g(t) \in \left[-2, \frac{9}{4}\right]$.

When $t = -1$ and $b = 1$, $f(x) = -2x + 1$, S achieves its minimum value -2.

When $t = 1$ and $b = \frac{1}{2}$, $f(x) = \frac{1}{2}x + \frac{1}{2}$, S achieves its maximum value $\frac{9}{4}$.

Thus, the range of S is $\left[-2, \frac{9}{4}\right]$. $\square$

6   In $\triangle ABC$, $BC = a$, $CA = b$, $AB = c$, and $c < b < a < 2c$. Two points P and Q are on the boundary of $\triangle ABC$, and the line PQ divides $\triangle ABC$ into two parts with the same area. Determine the minimum length of the segment PQ.

Solution First consider the situation when P and Q are on AB and AC. Without loss of generality, P is on AB and Q is on AC.

Let $AP = x$ and $AQ = y$. Since the area of $\triangle APQ$ is half of the area of $\triangle ABC$, $xy \sin A = \dfrac{1}{2} bc \sin A$; so $xy = \dfrac{1}{2} bc$.

By the law of cosines,

$$|PQ|^2 = x^2 + y^2 - 2xy \cos A$$
$$\geq 2xy - 2xy \cos A$$
$$= bc(1 - \cos A)$$
$$= 2bc \sin^2 \frac{A}{2}.$$

Since $c < b < a < 2c$, we have $\sqrt{\dfrac{bc}{2}} < c < b$. So, when $x = y = \sqrt{\dfrac{bc}{2}}$, P and Q are indeed on the segments AB and AC, respectively, and PQ^2 achieves its minimum

$$d(a) = 2bc \sin^2 \frac{A}{2}.$$

Since $a = 2R \sin A$, where R is the radius of the circumcircle of $\triangle ABC$, we have

$$d(a) = 2bc \sin^2 \frac{A}{2} = \frac{abc}{R \sin A} \sin^2 \frac{A}{2} = \frac{abc}{2R} \tan \frac{A}{2}.$$

From $c < b < a < 2c$, we also have $\sqrt{\dfrac{ca}{2}} < c < a$ and $\sqrt{\dfrac{ab}{2}} < b < a$. So, similarly, when P and Q are on the sides AB and BC, the minimum value of PQ^2 is

$$d(b) = \frac{abc}{2R} \tan \frac{B}{2};$$

and when P and Q are on the sides AC and BC, the minimum value of PQ^2 is

$$d(c) = \frac{abc}{2R} \tan \frac{C}{2}.$$

Since the tangent function is increasing on $\left(0, \dfrac{\pi}{2}\right)$, and $0 < C < B < A < \pi$, we have

$$d(c) < d(b) < d(a).$$

So the minimum length of PQ is $\sqrt{d(c)} = \sqrt{2ab} \sin \dfrac{C}{2}$. □

Remark. The answer can also be expressed in other forms. For example, $\sqrt{\dfrac{c^2 - (a-b)^2}{2}}$, or $\sqrt{\dfrac{1}{2}(a+c-b)(b+c-a)}$.

7 As in the figure, in the $\triangle ABC$, $AB > AC > BC$; the incircle of $\triangle ABC$ touches the sides AB, BC, and CA at points D, E, and F, respectively; L, M, and N are the midpoints of DE, EF, and FD, respectively. The line NL intersects the ray AB at P; the line LM intersects the ray BC at Q; and the line NM intersects the ray AC at R.

Prove that $PA \cdot QB \cdot RC = PD \cdot QE \cdot RF$.

Proof. As in Figure 7.1, let S be the intersection of the lines DE and AR. By Menelaus' theorem,

$$\frac{AD}{DB} \cdot \frac{BE}{EC} \cdot \frac{CS}{SA} = 1.$$

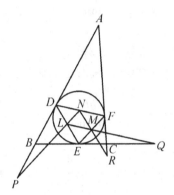

Fig. 7.1

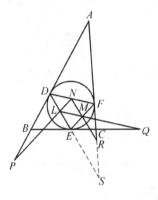

Fig. 7.2

Since $AD = AF$, $DB = BE$, and $EC = CF$, so

$$\frac{CS}{CF} = \frac{SA}{AF}. \qquad \qquad ①$$

Since M and N are midpoints of EF and FD, so R is the midpoint of FS, therefore

$$CS - CF = 2RC, \quad SA - AF = 2RF.$$

These, together with $①$, imply

$$\frac{2RC}{CF} = \frac{CS - CF}{CF} = \frac{SA - AF}{AF} = \frac{2RF}{CF},$$

i.e.,

$$\frac{RC}{CF} = \frac{RF}{AF}.$$

Therefore,

$$\frac{RC}{RF} = \frac{RC}{RC + CF} = \frac{RF}{RF + AF} = \frac{RF}{RA},$$

i.e.,

$$RF^2 = RC \cdot RA.$$

Similarly,

$$PD^2 = PA \cdot PB, \quad QE^2 = QB \cdot QC.$$

So each of P, Q, and R has the same power to the circumcircle and incircle of $\triangle ABC$, therefore they are collinear.

By Menelaus' theorem,

$$\frac{AP}{PB} \cdot \frac{BQ}{QC} \cdot \frac{CR}{RA} = 1,$$

so

$$\frac{AP^2}{PD^2} \cdot \frac{BQ^2}{QE^2} \cdot \frac{CR^2}{RF^2} = \frac{AP^2}{PA \cdot PB} \cdot \frac{BQ^2}{QB \cdot QC} \cdot \frac{CR^2}{RC \cdot RA}$$

$$= \frac{AP}{PB} \cdot \frac{BQ}{QC} \cdot \frac{CR}{RA} = 1,$$

that is

$$PA \cdot QB \cdot RC = PD \cdot QE \cdot RF. \qquad \qquad \square$$

8 Given integers m and n, let $A(m,n) = \{x^2 + mx + n : x \in \mathbb{Z}\}$, where $\mathbb{Z}$ is the set of integers. Is it always true that there are three distinct integers $a, b, c \in A(m, n)$ so that $a = bc$? Justify your answer.

Solution   We first prove that, for any integer n, $A(0, n)$ and $A(1, n)$ has the property in the statement.

Indeed, pick integer r big enough, so that $0 < r < r + 1 < n + r(r + 1)$, and let

$$a = (n + r(r + 1))^2 + n,$$
$$b = r^2 + n,$$
$$c = (r + 1)^2 + n.$$

Then, $a, b, c \in A(0, n)$, $b < c < a$, and

$$a = n^2 + (2r(r + 1) + 1)n + r^2(r + 1)^2$$
$$= (n + r^2)(n + (r + 1)^2)$$
$$= bc.$$

So $A(0, n)$ has the property in the statement.

Similarly, pick integer r big enough so that $0 < r - 1 < r < n + r^2 - 1$, and let

$$a = (n + r^2 - 1)(n + r^2) + n,$$
$$b = (r - 1)r + n,$$
$$c = r(r + 1) + n.$$

Then, $a, b, c \in A(1, n)$, $b < c < a$, and

$$a = n^2 + 2r^2 n + (r^2 - 1)r^2$$
$$= (n + r(r - 1))(n + r(r + 1))$$
$$= bc.$$

So $A(1, n)$ has the property in the statement, too.

Now we prove that, for any integers k and n, the sets $A(2k, n)$ and $A(2k + 1, n)$ has the property in the statement. In fact, since x ranges over all the integers if and only if $x_1 = x + k$ ranges over all the integers, and

$$x^2 + 2kx + n = (x + k)^2 - k^2 + n = x_1^2 + (n - k^2),$$

$$x^2 + (2k+1)x + n = (x+k)(x+k+1) - k(k+1) + n = x_1^2 + x_1 + (n - k^2 - k),$$

so

$$A(2k, n) = A(0, n - k^2),$$
$$A(2k + 1, n) = A(1, n - k^2 - k).$$

Therefore, $A(2k, n)$ and $A(2k + 1, n)$ has the property in the statement.

In conclusion, for any integers m and n, there are three distinct integers a, b, and c in $A(m, n)$ such that $a = bc$. $\qquad \square$

11th Grade

First Day

July 27, 2015

8:00–12:00

1. In $\triangle ABC$, $AB > AC$; I is its incenter. The circle Γ has AI as its diameter and intersects the circumcircle of $\triangle ABC$ at A amd D, where D is on the arc $\widehat{AC}$ that does not contain B. The line passing through A and parallel to BC intersects Γ at A and another point E, as in Figure 1.1. Suppose DI bisects $\angle CDE$, and $\angle ABC = 33°$. Determine the degree of $\angle BAC$.

The problem is identical to Problem 2, 10th grade; so is the solution.

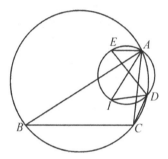

Fig. 1.1

2. A series $\{a_n\}$ satisfies $a_1 = 1$, $a_{2k} = a_{2k-1} + a_k$, and $a_{2k+1} = a_{2k}$ for all $k = 1, 2, \ldots$.
   Prove that

$$a_{2^n} > 2^{\frac{n^2}{4}}$$

for any positive integer n.

Proof. It is easy to see that $\{a_n\}$ is monotone non-decreasing.
   Moreover, we prove that

$$a_{s+t+1} + a_{s-t} \geq a_{s+t} + a_{s-t+1} \tag{1}$$

for all integers s and t, where $0 \leq t < s$.

Indeed, when s and t are of the same parity, $s \pm t$ are even, and we have $a_{s+t+1} = a_{s+t}$, $a_{s-t} = a_{s-t+1}$, so the two sides of (1) are actually equal, and (1) holds.

When s and t are of different parity, $s \pm t + 1$ are even, so

$$(a_{s+t+1} + a_{s-t}) - (a_{s+t} + a_{s-t+1}) = (a_{s+t+1} - a_{s+t}) - (a_{s-t+1} - a_{s-t})$$

$$= a_{\frac{s+t+1}{2}} - a_{\frac{s-t+1}{2}} \geq 0.$$

So, in any case, (1) holds.

By (1), $a_{2s} + a_1 \geq a_{2s-1} + a_2 \geq \cdots \geq a_{s+1} + a_s$, so

$$a_1 + a_2 + \cdots + a_{2s} \geq s(a_{s+1} + a_s) \geq 2sa_s.$$

On the other hand,

$$a_1 + a_2 + \cdots + a_{2s} = \sum_{i=1}^{2s}(a_{2i} - a_{2i-1})$$

$$= a_{4s} - \left(\sum_{i=1}^{2s-1}(a_{2i+1} - a_{2i})\right) - a_1$$

$$< a_{4s}.$$

So we have $a_{4s} > 2sa_{2s}$, in particular, when $m \geq 2$ and $s = 2^{m-2}$,

$$a_{2^m} > 2^{m-1}a_{2^{m-2}}. \qquad (2)$$

Now we prove by induction on n that $a_{2^n} > 2^{\frac{n^2}{4}}$ for any positive integer n.

When $n = 1$, $a_2 = 2 > 2^{\frac{1^2}{4}}$; when $n = 2$, $a_4 = 4 > 2^{\frac{2^2}{4}}$.

Assume the statement holds for $n = m - 2$, by (2) and the inductive hypothesis,

$$a_{2^m} > 2^{m-1}a_{2^{m-2}} > 2^{m-1+\frac{(m-2)^2}{4}} = 2^{\frac{m^2}{4}}.$$

So the statement holds when $n = m$. Thus, for all positive integers n, $a_{2^n} > 2^{\frac{n^2}{4}}$. $\qquad \square$

3 For each positive integer n, define $P_n = \{n^k : k = 0, 1, 2, \ldots\}$. We call a triple (a, b, c) of positive integers *lucky* if there is a positive integer m such that the three (not necessarily distinct) numbers $a - 1$, $ab - 12$, and $abc - 2015$ all belong to P_m.

Determine the number of lucky triples.

The problem is identical to Problem 4, 10th grade; so is the solution.

4 $a_1, a_2, \ldots, a_8$ are 8 distinct positive integers; any three of them has greatest common divisor 1.

Prove that there is an integer $n \geq 8$, and n distinct positive integers $m_1, m_2, \ldots, m_n$, such that the greatest common divisor of $m_1, m_2, \ldots,$ and m_n is 1, and, for any integers p, q, and r, where $1 \leq p < q < r \leq n$, there exist i and j such that $1 \leq i < j \leq 8$ and $a_i a_j \mid (m_p + m_q + m_r)$.

Proof. We may assume $a_1 < a_2 < \cdots < a_8$. Let M be the set of integers consisting of the products of any six distinct a_i's.

$|M| \geq 8$ because it contains, from the smallest to the largest, the 8 numbers $a_1 a_2 a_3 a_4 a_5 a_6$, $a_1 a_2 a_3 a_4 a_5 a_7$, $a_1 a_2 a_3 a_4 a_5 a_8$, $a_1 a_2 a_3 a_4 a_6 a_8$, $a_1 a_2 a_3 a_4 a_7 a_8$, $a_1 a_2 a_3 a_5 a_7 a_8$, $a_1 a_2 a_3 a_6 a_7 a_8$, and $a_1 a_2 a_4 a_6 a_7 a_8$.

Let $n = |M| \geq 8$, and $m_1, m_2, \ldots, m_n$ are the elements of M.

We first prove that the greatest common divisor of $m_1, m_2, \ldots,$ and m_n is 1.

Indeed, consider any prime p. Since any three of the a_i's has greatest common divisor 1, so p divides at most two of the a_i's and does not divide at least six of them, so p does not divide the product of those six, which is an element of M. So, the greatest common divisor of $m_1, m_2, \ldots,$ and m_n is 1. Now consider m_p, m_q, and m_r for any $1 \leq p < q < r \leq n$. By the construction of M, there are 6-element subsets A_p, A_q, and A_r of $A = \{1, 2, \ldots, 8\}$ such that

$$m_p = \prod_{u \in A_p} a_u, \quad m_q = \prod_{v \in A_q} a_v, \quad m_r = \prod_{w \in A_r} a_w. \qquad ①$$

Since

$$
\begin{aligned}
8 - |A_p \cap A_q \cap A_r| &= |A \setminus (A_p \cap A_q \cap A_r)| \\
&= |(A \setminus A_p) \cup (A \setminus A_q) \cup (A \setminus A_r)| \\
&\leq |A \setminus A_p| + |A \setminus A_q| + |A \setminus A_r| \\
&= 2 + 2 + 2 = 6,
\end{aligned}
$$

so there are two numbers i and j in $A_p \cap A_q \cap A_r$. By ①, $a_i a_j \mid m_p$, $a_i a_j \mid m_q$, and $a_i a_j \mid m_r$; so

$$a_i a_j \mid (m_p + m_q + m_r).$$

In conclusion, n and $m_1, m_2, \ldots, m_n$ are the numbers satisfying our requirement. □

Second Day

July 28, 2015

8:00–12:00

5   As in Figure 5.1, E and F are points on the segments AB and AD, respectively; BF and DE intersect at the point C; $AE + EC = AF + FC$.

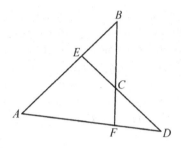

Fig. 5.1

Prove that $AB + BC = AD + DC$.

Proof. As in Figure 5.2, draw the segment AC. Denote $AC = a$, $\angle EAC = \alpha$, $\angle CAD = \beta$, $\angle ABC = \gamma$, and $\angle ADC = \theta$.

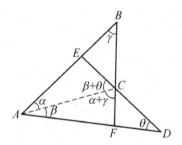

Fig. 5.2

By the law of sines,

$$AE + EC = a \cdot \frac{\sin \alpha + \sin(\beta + \theta)}{\sin(\alpha + \beta + \theta)},$$

$$AF + FC = a \cdot \frac{\sin \beta + \sin(\alpha + \gamma)}{\sin(\beta + \alpha + \gamma)}.$$

So the condition that $AE + EC = AF + FC$ is equivalent to

$$\frac{\sin \alpha + \sin(\beta + \theta)}{\sin(\alpha + \beta + \theta)} = \frac{\sin \beta + \sin(\alpha + \gamma)}{\sin(\beta + \alpha + \gamma)},$$

which in turn, by the sum-to-product identity and the double-angle formula, is equivalent to

$$\frac{\cos \dfrac{\beta + \theta - \alpha}{2}}{\cos \dfrac{\beta + \theta + \alpha}{2}} = \frac{\cos \dfrac{\alpha + \gamma - \beta}{2}}{\cos \dfrac{\alpha + \gamma + \beta}{2}},$$

i.e.,

$$\cos \frac{\beta + \theta - \alpha}{2} \cos \frac{\alpha + \gamma + \beta}{2} = \cos \frac{\beta + \theta + \alpha}{2} \cos \frac{\alpha + \gamma - \beta}{2},$$

which, by the product-to-sum identity, is equivalent to

$$\cos \left(\beta + \frac{\theta}{2} + \frac{\gamma}{2} \right) + \cos \left(\alpha + \frac{\gamma}{2} - \frac{\theta}{2} \right)$$

$$= \cos \left(\alpha + \frac{\gamma}{2} + \frac{\theta}{2} \right) + \cos \left(\beta + \frac{\theta}{2} - \frac{\gamma}{2} \right) \qquad \text{(1)}$$

By the law of sines,

$$AD + DC = a \cdot \frac{\sin \beta + \sin(\beta + \theta)}{\sin \theta},$$

$$AB + BC = a \cdot \frac{\sin \alpha + \sin(\alpha + \gamma)}{\sin \gamma}.$$

So we only need to show

$$\frac{\sin \beta + \sin(\beta + \theta)}{\sin \theta} = \frac{\sin \alpha + \sin(\alpha + \gamma)}{\sin \gamma},$$

which, by the sum-to-product identity and the double-angle formula, is equivalent to

$$\frac{\sin \left(\beta + \dfrac{\theta}{2} \right)}{\sin \dfrac{\theta}{2}} = \frac{\sin \left(\alpha + \dfrac{\gamma}{2} \right)}{\sin \dfrac{\gamma}{2}},$$

i.e.,

$$\sin \left(\beta + \frac{\theta}{2} \right) \sin \frac{\gamma}{2} = \sin \left(\alpha + \frac{\gamma}{2} \right) \sin \frac{\theta}{2}.$$

which, by the product-to-sum identity, is equivalent to

$$\cos\left(\beta + \frac{\theta}{2} - \frac{\gamma}{2}\right) - \cos\left(\beta + \frac{\theta}{2} + \frac{\gamma}{2}\right)$$
$$= \cos\left(\alpha + \frac{\gamma}{2} - \frac{\theta}{2}\right) - \cos\left(\alpha + \frac{\gamma}{2} + \frac{\theta}{2}\right). \qquad ②$$

Clearly ① and ② are equivalent.   □

6 Given integer $n \geq 2$, and let $A = \{(a, b) : a, b \in \{1, 2, \ldots, n\}\}$. Colour each point in A with one of red, yellow, and blue, under the restriction that $(a, b+1)$ and $(a+1, b+1)$ have the same colour whenever (a, b) and $(a+1, b)$ have the same colour. Determine the total number of such colourings.

Solution   For $k = 1, 2, \ldots, n$, define the set of points

$$A_k = \{(k, b) : b \in \{1, 2, \ldots, n\}\}.$$

We successively colour the sets A_1, A_2, $\ldots$, A_n, and denote by N_k, $k = 1, 2, \ldots, n$, the number of colourings of A_k in our process.

First we colour A_1. There are 3 choices for each point $(1, b)$, $b = 1, 2, \ldots, n$; $N_1 = 3^n$.

Suppose we finished the colouring of $A_1 \cup \cdots \cup A_r$, where $1 \leq r \leq n-1$; now we colour A_{r+1}.

If the colours on $(r+1, s)$ and (r, s) are different for every $s = 1, 2, \ldots, n$, we have 2 choices for each point in A_{r+1}, and there are 2^n such colourings.

Otherwise, let $s \in \{1, 2, \ldots, n\}$ be the smallest number such that $(r+1, s)$ and (r, s) have the same colour. By our rule, we must colour $(r+1, v)$ with the same colour as (r, v) for each $v = s, s+1, \ldots, n$. However, for each u, where $1 \leq u < s$, we can colour $(r+1, u)$ with any of the 2 colours that are not used for (r, u). So in such a situation, there are 2^{s-1} colourings. Note that this is also true for $s = 1$.

Therefore, no matter how we coloured $A_1 \cup \cdots \cup A_r$, the number of choices for A_{r+1} is

$$N_{r+1} = 2^n + \sum_{s=1}^{n} 2^{s-1} = 2^{n+1} - 1.$$

By the rule of product, the total number of colourings is

$$N_1 N_2 \ldots N_n = 3^n \cdot (2^{n+1} - 1)^{n-1}.$$

□

7 As in Figure 7.1, in the $\triangle ABC$, $AB > AC > BC$; the incircle of $\triangle ABC$ touches the sides AB, BC, and CA at points D, E, and F, respectively; L, M, and N are the midpoints of DE, EF, and FD, respectively. The line NL intersects the ray AB at P; the line LM intersects the ray BC at Q; and the line NM intersects the ray AC at R.

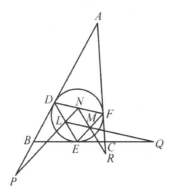

Fig. 7.1

Prove that $PA \cdot QB \cdot RC = PD \cdot QE \cdot RF$.

The problem is identical to Problem 7, 10th grade; so is the solution.

8 Determine all the prime numbers p such that there exists an integer coefficient polynomial $f(x) = x^{p-1} + a_{p-2}x^{p-2} + \ldots + a_1 x + a_0$ satisfying: $f(x)$ has $p - 1$ consecutive positive integer as its roots, and $p^2 \mid f(\mathrm{i})f(-\mathrm{i})$, where i is the imaginary unit.

Proof. Suppose f exists and its $p-1$ roots are $m, m+1, \ldots, m+p-2$, so

$$f(x) = (x - m)(x - m - 1) \ldots (x - m - p + 2),$$

and

$$f(\mathrm{i})f(-\mathrm{i}) = \prod_{k=m}^{m+p-2} (\mathrm{i} - k) \cdot \prod_{k=m}^{m+p-2} (-\mathrm{i} - k) = \prod_{k=m}^{m+p-2} (k^2 + 1).$$

Since $p^2 \mid f(\mathrm{i})f(-\mathrm{i})$, we have

$$p^2 \mid \prod_{k=m}^{m+p-2} (k^2 + 1). \qquad \qquad \textcircled{1}$$

When $p = 2$, (1) becomes $4 \mid (m^2 + 1)$, but this is not possible for any integer m; (1) does not hold.

When $p \equiv 3 \pmod 4$, -1 is not a quadratic residue of p, so there is no k where $k^2 + 1 \equiv 0 \pmod p$; therefore, the right hand side of (1) is not a multiple of p, (1) never holds.

When $p \equiv 1 \pmod 4$, -1 is a quadratic residue of p, and there exists $k_1 \in \{1, 2, \ldots, p-1\}$ so that $p \mid (k_1^2 + 1)$. Note that $k_2 = p - k_1$ also satisfies $k_2 \in \{1, 2, \ldots, p-1\}$ and $p \mid (k_2^2 + 1)$; moreover $k_2 \neq k_1$. So $p^2 \mid \prod_{k=1}^{p-1}(k^2 + 1)$ and (1) holds when $m = 1$.

In conclusion, the primes we seek are those p such that $p \equiv 1 \pmod 4$, and we can take

$$f(x) = (x - 1)(x - 2) \ldots (x - p + 1)$$

as the corresponding polynomial. $\qquad\square$

International Mathematical Olympiad

2015 (Chiang Mai, Thailand)

First Day

July 10, 2015

9:00–13:30

1 We say that a finite set S in the plane is *balanced* if, for any two different points A and B in S, there is a point C in S such that $AC = BC$. We say that S is *centre-free* if for any three points A, B and C in S, there is no point P in S such that $PA = PB = PC$.

(a) Show that for all integers $n \geq 3$, there exists a balanced set consisting of n points.

(b) Determine all integers $n \geq 3$ for which there exists a balanced centre-free set consisting of n points. (posed by Netherlands)

Proof. (a) For an odd number $n \geq 3$, let S be the set of n vertices of a regular n-gon. We show that S is balanced. Indeed, the points of S are distributed evenly on a circle ω. For any two points $A, B \in S$, they divide ω into two arcs, one of which has an odd number of points in S; the midpoint C of this arc is in S, and $AC = BC$.

For an even number $n \geq 4$, consider the following: Let ω be a circle with center O. Let $k = n/2 - 1$ and pick $A_1, A_2, \ldots, A_k$ be k points on ω that are very close to each other. To be specific, these k points lie on an arc with central angle $30°$. Rotate each of these points around O $60°$ degrees clockwise and we get k points $B_1, B_2, \ldots, B_k$. And rotate A_1 around O counter-clockwise and we get A'. Let

$$S = \{O, A_1, \ldots, A_k, B_1, \ldots, B_k, A'\}.$$

S has n distinct points. And we prove that S is balanced. Let A and B be any two distinct points in S. When both A and B are on ω, then $OA = OB$; otherwise, one of A and B is O, by our construction, there is another point $C \in S$ such that $\triangle ABC$ is equilateral, and $CA = CB$.

(b) The answer is all odd numbers $n \geq 3$.

When $n \geq 3$ is an odd number, let S be the set of n vertices of a regular n-gon. We proved in (a) that S is balanced. The circumcenter of any three points in S is the center of the polygon, which is not in S; so S is centre-free.

For any even number $n \geq 4$, we prove that there is no centre-free set with n points. Suppose S is a balanced set with n points. For any subset $\{A, B\} \subseteq S$, there are points in S that are equidistant from A and B, we pick any of such points and call it the *connection point* of $\{A, B\}$. There are $\binom{n}{2}$ binary subsets of S, each determine a connection point. By the pigeonhole principle, there is a point $P \in S$ which is the connection point for at least $\dfrac{1}{n}\binom{n}{2} = \dfrac{1}{2}(n - 1)$ binary subsets of S.

Note that n is odd, so P is the connection point for at least $n/2$ binary subsets. P is in none of these binary sets, so their elements are from $S \backslash \{P\}$.

Since $(n/2) \times 2 = n > n - 1$, two of these binary subsets overlap, say, $\{A, B\}$ and $\{A, C\}$. Hence $PA = PB = PC$, and S is not centre-free. $\quad\square$

2 Determine all triples (a, b, c) of positive integers such that each of the numbers

$$ab - c, \quad bc - a, \quad ca - b$$

is a power of 2.

(A power of 2 is an integer of the form 2^n, where n is a non-negative integer). (posed by Serbia)

Solution   Let (a, b, c) be such a triple. $a = 1$ would imply that both $b - c$ and $c - b$ are powers of 2, which is impossible. So $a \geq 2$; similarly, $b \geq 2$ and $c \geq 2$. We discuss two cases.

Case 1. At least two of a, b, and c are the same.

Without loss of generality, $a = b$. Then $ac - b = a(c - 1)$ is a power of 2, so are both a and $c - 1$. Let $a = 2^s$ and $c = 1 + 2^t$, where $s \geq 1$ and

$t \geq 0$. $ab - c = 2^{2s} - 2^t - 1$ is a power of 2. When $t > 0$, $2^{2s} - 2^t - 1$ is an odd number, so $2^{2s} - 2^t - 1 = 1$. Then we have $2^t \equiv 2 \pmod 4$, so $t = 1$ and $s = 1$, $a = b = 2$ and $c = 3$. When $t = 0$, $2^{2s} - 2$ is a power of 2, we must have $s = 1$ and $a = b = c = 2$. It is easy to verify that both triples $(2, 2, 2)$ and $(2, 2, 3)$ satisfy our requirement.

Case 2. a, b, and c are pairwise distinct.

Without loss of generality, $2 \leq a < b < c$. By our assumption, there are non-negative integers α, β, and γ such that

$$bc - a = 2^\alpha, \qquad\qquad \text{①}$$

$$ca - b = 2^\beta, \qquad\qquad \text{②}$$

$$ab - c = 2^\gamma. \qquad\qquad \text{③}$$

It is easy to see that $\alpha > \beta > \gamma \geq 0$. We discuss two sub-cases based on the value of a.

Case 2.1. $a = 2$.

We first prove that $\gamma = 0$. Otherwise, by ③, c is even; and by ②, b is even; thus, the left hand side of ①, $bc - a \equiv 2 \pmod 4$; yet $2^\alpha \equiv 0 \pmod 4$, a contradiction.

So $\gamma = 0$. ③ becomes $c = 2b - 1$, combined with ② we get $3b - 2 = 2^\beta$. Take modulo 3 we get that β is even. If $\beta = 2$, $b = 2 = a$ contradicts our assumption in this case. If $\beta = 4$, we get $b = 6$ and $c = 11$; and it is easy to verify that $(2, 6, 11)$ is a triple that satisfy our requirement. If $\beta \geq 6$,

$$b = \frac{1}{3}(2^\beta + 2).$$

Substitute into ①,

$$9 \cdot 2^\alpha = 9(bc - a) = 9b(2b - 1) - 18$$

$$= (3b - 2)(6b + 1) - 16$$

$$= 2^\beta (2^{\beta+1} + 5) - 16.$$

Since $\alpha > \beta \geq 6$, $2^7 \mid 9 \cdot 2^\alpha$; however the right hand side of the above only divides 2^4 but not 2^5, a contradiction.

Case 2.2. $a \geq 3$.

Adding ① and ②, we get

$$(a + b)(c - 1) = 2^\alpha + 2^\beta.$$

Subtracting ② from ①, we get

$$(b - a)(c + 1) = 2^\alpha - 2^\beta.$$

One of $c+1$ and $c-1$ is not a multiple of 4, so $2^{\beta-1} \mid a+b$ or $2^{\beta-1} \mid b-a$. From

$$2^\beta = ac - b \geq 3c - b > 2c$$

we get $b < c < 2^{\beta-1}$, so $0 < b-a < 2^{\beta-1}$, and it is impossible for $2^{\beta-1} \mid b-a$. Therefore $2^{\beta-1} \mid a+b$, and since $a+b < 2b < 2^\beta$, we must have $a+b = 2^{\beta-1}$. Substitute into ②,

$$ac - b = 2^\beta = 2(a+b),$$

i.e., $a(c-2) = 3b$. When $a \geq 4$, we have $b \geq 5$,

$$a(c-2) \geq 4(c-2) \geq 4(b-1) > 3b,$$

a contradiction. So $a = 3$, $c - 2 = b$, therefore $b = 2^{\beta-1} - 3$, $c = 2^{\beta-1} - 1$. Substitute into ③,

$$2^\gamma = ab - c = 3(2^{\beta-1} - 3) - (2^{\beta-1} - 1) = 2^\beta - 8,$$

which implies $\beta = 4$, $b = 5$, $c = 7$. It is easy to verify that $(3, 5, 7)$ satisfies our requirement.

So there are 16 such triples — $(2, 2, 2)$, the 3 permutations of $(2, 2, 3)$, the 6 permutations of $(2, 6, 11)$, and the 6 permutations of $(3, 5, 7)$. □

3 Let ABC be an acute triangle with $AB > AC$. Let Γ be its circumcircle, H its orthocenter, and F the foot of the altitude from A. Let M be the midpoint of BC. Let Q be the point on Γ such that $\angle HQA = 90°$, and let K be the point on Γ such that $\angle HKQ = 90°$. Assume that the points A, B, C, K, and Q are all different, and lie on Γ in this order.

Prove that the circumcircles of triangles KQH and FKM are tangent to each other. (posed by Ukraine)

Proof. As in Figure 3.1, extend QH meet Γ at A'. Since $\angle AQH = 90°$, AA' is a diameter of Γ. Since $A'B \perp AB$, so $A'B \parallel CH$. Similarly, $A'C \parallel BH$, so $BA'CH$ is a parallelogram, and M is the midpoint of $A'H$. Extend AF meet Γ at E, since $A'E \perp AE$, so $A'E \parallel BC$, and MF is a midline of $\triangle HA'E$, F is the midpoint of HE.

Let R be the intersection of $A'E$ and QK. By the power of a point theorem,

$$RK \cdot RQ = RE \cdot RA'.$$

ω_1, the circumcircle of $\triangle HKQ$ and ω_2, the circumcircle of $\triangle HEA'$ are circles with diameters HQ and HA', respectively. They are externally tangent

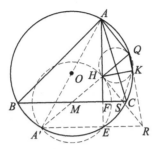

Fig. 3.1

at H, and the point R has the same power to them, so R is on the radical axis of ω_1 and ω_2; so RH is their common tangent line, and $RH \perp A'Q$.

Let S be the intersection of the lines MF and RH, then S is the midpoint of RH. Since $\triangle RHK$ is a right triangle, and S is the midpoint of its hypotenuse RH, so $SH = SK$. And because SH is tangent to ω_1, we get SK is also tangent to ω_1. In the right triangle SHM, HF is the altitude on the hypotenuse, so

$$SF \cdot SM = SH^2 = SK^2,$$

so SK is also tangent to the circumcircle of $\triangle KMF$. Therefore SK is tangent to both circumcircles of $\triangle KQH$ and $\triangle FKM$ at the point K, and these two circles are tangent at K as well. $\qquad\square$

Second Day

July 11, 2015

9:00–13:30

4 Triangle ABC has circumcircle Ω and circumcenter O. A circle Γ with center A intersects the segment BC at points D and E, such that B, D, E, and C are all different and lie on line BC in this order. Let F and G be the points of intersection of Γ and Ω, such that A, F, B, C, and G lie on Ω in this order. Let K be the second point of intersection of the circumcircle of triangle BDF and the segment AB. Let L be the second point of intersection of the circumcircle of triangle CGE and the segment CA.

Suppose that the lines FK and GL are different and intersect at the point X. Prove that X lies on the line AO. (posed by Greece)

Proof. As in Figure 4.1, since $AF = AG$, and AO is the bisector of $\angle FAG$, so F and G are symmetric about the line AO. In order to prove that X is on AO, it is enough to show that $\angle AFK = \angle AGL$.

First of all,

$$\angle AFK = \angle DFG + \angle GFA - \angle DFK.$$

Since D, F, G, E are concyclic, we have $\angle DFG = \angle CEG$. Because A, F, B, G are concyclic, $\angle GFA = \angle GBA$. And because D, B, F, K are

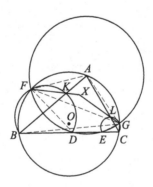

Fig. 4.1

concyclic, $\angle DFK = \angle DBK$. So,

$$\angle AFK = \angle CEG + \angle GBA - \angle DBK = \angle CEG - \angle CBG.$$

Since C, E, L, G are concyclic, $\angle CEG = \angle CLG$; and C, B, A, G are concyclic, so $\angle CBG = \angle CAG$. So,

$$\angle AFK = \angle CLG - \angle CAG = \angle AGL. \qquad \square$$

5 Let $\mathbb{R}$ be the set of real numbers. Determine all functions $f : \mathbb{R} \to \mathbb{R}$ satisfying the equation

$$f(x + f(x + y)) + f(xy) = x + f(x + y) + yf(x)$$

for all real numbers x and y. (posed by Albania)

Solution　Denote the equation in the problem as $P(x, y)$, and let the function f be a solution to our problem. Consider $P(x, 1)$,

$$f(x + f(x + 1)) = x + f(x + 1). \qquad \text{①}$$

Therefore, for any real number x, $x + f(x + 1)$ is a fixed point of f. We discuss two cases.

Case 1. $f(0) \neq 0$.
Consider $P(0, y)$,

$$f(f(y)) + f(0) = f(y) + yf(0).$$

Suppose y_0 is a fixed point of f, set $y = y_0$ in the last equation, we have $y_0 = 1$. So,

$$x + f(x + 1) = 1,$$

and $f(x) = 2 - x$ for all x. It is easy to verify that $f(x) = 2 - x$ is a solution to our problem.

Case 2. $f(0) = 0$.
Observe $P(x + 1, 0)$,

$$f(x + f(x + 1) + 1) = x + f(x + 1) + 1. \qquad \text{②}$$

And observe $P(1, y)$,

$$f(1 + f(y + 1)) + f(y) = 1 + f(y + 1) + yf(1). \qquad \text{③}$$

Set $x = -1$ in ① we get $f(-1) = -1$, and set $y = -1$ in ③ we get $f(1) = 1$. Therefore we can rewrite ③ as

$$f(1 + f(y + 1)) + f(y) = 1 + f(y + 1) + y. \qquad \text{④}$$

Suppose both y_0 and $y_0 + 1$ are fixed points of f, set $y = y_0$ in ④ we get $y_0 + 2$ as a fixed point as well. So, by ① and ②, for any real number x, $x + f(x + 1) + 2$ is a fixed point of f, i.e.,

$$f(x + f(x + 1) + 2) = x + f(x + 1) + 2.$$

Replace x by $x - 2$ in the last equation,

$$f(x + f(x - 1)) = x + f(x - 1).$$

Observe $P(x, -1)$,

$$f(x + f(x - 1)) = x + f(x - 1) - f(x) - f(-x).$$

By the last two equations, $f(-x) = -f(x)$, i.e., f is an odd function. Observe $P(-1, -y)$, and note that $f(-1) = -1$, we have

$$f(-1 + f(-y - 1)) + f(y) = -1 + f(-y - 1) + y.$$

And by the fact that f is odd, the last equation can be written as

$$-f(1 + f(y + 1) + f(y)) = -1 - f(y + 1) + y.$$

Add this with ④, $f(y) = y$ for all real y. It is easy to verify that $f(x) = x$ is a solution to our problem.

In conclusion, there are two such functions $f(x) = x$ and $f(x) = 2 - x$.

□

6 The sequence $a_1, a_2, \ldots$ of integers satisfies the conditions:

(i) $1 \le a_j \le 2015$ for all $j \ge 1$;
(ii) $k + a_k \ne \ell + a_\ell$ for all $1 \le k < \ell$.

Prove that there exist two positive integers b and N such that

$$\left| \sum_{j=m+1}^{n} (a_j - b) \right| \le 1007^2$$

for all integers m and n such that $n > m \ge N$. (posed by Australia)

Proof. Let $s_n = n + a_n$, then $n + 1 \le s_n \le n + 2015$, and all the s_i's are distinct.

Let $S = \{s_1, s_2, \ldots\}$. We first prove that $M = \mathbb{N}^* \setminus S$ is finite, and $1 \le |M| \le 2015$.

Clearly $1 \in M$. Assume that there are more than 2015 elements in M, say, $m_1 < m_2 < \cdots < m_{2016}$. Pick an integer $n > m_{2016}$, then

$$\{s_1, s_2, \ldots, s_n\} \subseteq \{1, 2, \ldots, n + 2015\},$$

$$\{m_1, m_2, \ldots, m_{2016}\} \subseteq \{1, 2, \ldots, n + 2015\}.$$

$\{s_1, s_2, \ldots, s_n\}$ and $\{m_1, m_2, \ldots, m_{2016}\}$ are disjoint by the definition of M. However, $n + 2016 > n + 2015$, a contradiction. So M is finite and $1 \leq |M| \leq 2015$.

Let $b = |M|$, and pick an integer $N > \max M$, we prove that these numbers satisfy the requirement in the problem.

Clearly b and N are well defined integers, and we proved that $1 \leq b \leq 2015$. For any $n \geq N$, we have the following partition

$$\{1, 2, \ldots, n + 2015\} = \{s_1, s_2, \ldots, s_n\} \cup M \cup C_n, \qquad \text{①}$$

where $C_n = \{1, 2, \ldots, n+2015\} \backslash (\{s_1, s_2, \ldots, s_n\} \cup M)$, and $|C_n| = 2015 - b$. For $j \geq n + 1$, $s_j = j + a_j \geq n + 1 + 1 = n + 2$. If C_n is not empty, we must have $n + 2 \leq \min C_n$, otherwise, $\min C_n$ is not in S, nor in M, which is impossible. So

$$C_n \subseteq \{n + 2, n + 3, \ldots, n + 2015\}. \qquad \text{②}$$

Note that ② trivially holds when C_n is empty. Compute the summation of the elements on both sides of ①, writing $\sigma(X)$ the sum of the elements in a set X, we have

$$\sum_{i=1}^{n+2015} i = \sum_{j=1}^{n} s_j + \sigma(M) + \sigma(C_n).$$

Substitute s_j with $j + a_j$, we get

$$\sum_{i=n+1}^{n+2015} i = \sum_{j=1}^{n} a_j + \sigma(M) + \sigma(C_n).$$

For $n > m \geq N$, substitute n with m to get a new equation and take the difference,

$$\sum_{i=n+1}^{n+2015} i - \sum_{i=m+1}^{m+2015} i = \sum_{j=m+1}^{n} a_j + \sigma(C_n) - \sigma(C_m).$$

Subtract $(n-m)b$ from both sides and simplify, we get

$$\sum_{j=m+1}^{n} (a_j - b) = (2015 - b)(n - m) + \sigma(C_m) - \sigma(C_n). \qquad \text{(3)}$$

By (2) and $|C_n| = 2015 - b$ we have

$$\sum_{i=n+2}^{n+2016-b} i \le \sigma(C_n) \le \sum_{i=n+b+1}^{n+2015} i.$$

i.e.,

$$\left(n + 1009 - \frac{b}{2}\right)(2015 - b) \le \sigma(C_n) \le \left(n + 1008 + \frac{b}{2}\right)(2015 - b).$$

The same estimate holds with n replaced by m. Use these estimates in ·
(3) , we get

$$\sum_{j=m+1}^{n} (a_j - b) \le (2015 - b)(n - m)$$

$$+ \left(m + 1008 + \frac{b}{2}\right)(2015 - b) - \left(n + 1009 - \frac{b}{2}\right)(2015 - b)$$

$$= (2015 - b)(b - 1) \le 1007^2,$$

and

$$\sum_{j=m+1}^{n} (a_j - b) \ge (2015 - b)(n - m)$$

$$+ \left(m + 1009 - \frac{b}{2}\right)(2015 - b) - \left(n + 1008 + \frac{b}{2}\right)(2015 - b)$$

$$= -(2015 - b)(b - 1) \ge -1007^2.$$

Thus we have

$$\left| \sum_{j=m+1}^{n} (a_j - b) \right| \le 1007^2.$$

$\square$

International Mathematical Olympiad

2016 (Hongkong, China)

First Day

July 11, 2016

9:00–13:30

1. Triangle BCF has a right angle at B. Let A be the point on line CF such that $FA = FB$ and F lies between A and C. Point D is chosen so that $DA = DC$ and AC is the bisector of $\angle DAB$. Point E is chosen so that $EA = ED$ and AD is the bisector of $\angle EAC$. Let M be the midpoint of CF. Let X be the point such that $AMXE$ is a parallelogram (where $AM \parallel EX$ and $AE \parallel MX$). Prove that lines BD, FX, and ME are concurrent. (posed by Belgium)

Proof. As in Figure 1.1.

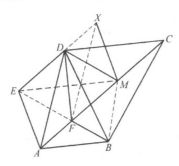

Fig. 1.1

We have,

$$\angle FAB = \angle FBA = \angle DAC = \angle DCA = \angle EAD = \angle EDA.$$

Denote their value by θ.

Because $\triangle ABF \sim \triangle ACD$, we have $AB : AC = AF : AD$, so $\triangle ABC \sim \triangle AFD$. And since $EA = ED$, we have

$$\angle AFD = \angle ABC = 90° + \theta = 180° - \frac{1}{2}\angle AED.$$

So F lies on the circle with center E and radius EA. In particular, $EF = EA = ED$. And note that

$$\angle EFA = \angle EAF = 2\theta = \angle BFC,$$

so B, F, and E are collinear.

Since $\angle EDA = \angle MAD$, we get $ED \parallel AM$, so E, D, and X are collinear.

M is the midpoint on the hypotenuse of the right triangle CBF, so $MF = MB$.

In the isosceles triangles EFA and MFB,

$$\angle EFA = \angle MFB, \quad AF = BF;$$

so they are congruent, and

$$BM = AE = XM,$$

and

$$BE = BF + FE = AF + FM = AM = EX.$$

Therefore $\triangle EMB \cong \triangle EMX$.

And since $EF = ED$, D and F are symmetric about the line EM; note that X and B are symmetric about EM as well; hence the lines BD and XF are symmetric about the line EM.

From this, we conclude that BD, FX, and ME are concurrent. $\square$

2 Find all positive integers n for which each cell of $n \times n$ table can be filled with one of the letters I, M and O in such a way that:

- in each row and each column, one third of the entries are I, one third are M and one third are O; and
- in any diagonal, if the number of entries on the diagonal is a multiple of three, then one third of the entries are I, one third are M and one third are O.

Note. The rows and columns of an $n \times n$ table are each labelled 1 to n in a natural order. Thus each cell corresponds to a pair of positive integer (i, j) with $1 \le i, j \le n$. For $n > 1$, the table has $4n - 2$ diagonals of two types. A diagonal of first type consists all cells (i, j) for which $i + j$ is a constant, and the diagonal of this second type consists all cells (i, j) for which $i - j$ is constant. (posed by Australia)

Solution The answer is all the multiples of 9.

We first give a 9×9 table as follows.

I	I	I	M	M	M	O	O	O
M	M	M	O	O	O	I	I	I
O	O	O	I	I	I	M	M	M
I	I	I	M	M	M	O	O	O
M	M	M	O	O	O	I	I	I
O	O	O	I	I	I	M	M	M
I	I	I	M	M	M	O	O	O
M	M	M	O	O	O	I	I	I
O	O	O	I	I	I	M	M	M

It is easy to verify that this table satisfies the requirement.

For $n = 9k$, where k is a positive integer, we first divide an $n \times n$ table into k^2 subtables of size 9×9, and fill each subtable with the above pattern. For the $n \times n$ table, each of its row, or column, or diagonal where the number of entries is a multiple of 3 intersects every subtable at a row, a column, or a diagonal with number of entries a multiple of 3, or 0 cells, in the subtable; so it has the same number of I, M, and O's.

Next we assume that an $n \times n$ table can be filled as required, and prove that $9 \mid n$. Since each row has the same number of I, M, and O's, so $3 \mid n$, and we denote $n = 3k$, where k is a positive integer. Divide the table into k^2 3×3 subtables, call the cell in the center of each subtable its *key cell*, and call a line — a row, a column, or a diagonal — passing through a key cell a *key line*. Let S be the set of all pairs (l, c), where l is a key line, and c is a cell on l that is filled with an M. We count the size of S in two ways.

On one hand, for each key line, exactly one third of its cells are filled with M. If l is one of the k key rows or k key columns, there are exactly k

cells on l filled with M. The key diagonals of each type has 3, 6, 9, ..., $3k$, $3k - 3$, ..., 3 cells on them, so

$$|S| = 2k \cdot k + 2 \cdot (1 + 2 + \cdots + k + (k - 1) + \cdots + 1)$$
$$= 2k^2 + 2k^2 = 4k^2.$$

On the other hand, for each cell c, it is incident to exactly 1 key line, if it is not a key cell, or 4 key lines, if it is a key cell. There are $3k^2$ cells filled with M, and $1 \equiv 4 \pmod{3}$, so

$$|S| \equiv 3k^2 \pmod{3}.$$

Therefore $4k^2 = |S| \equiv 3k^2 \pmod{3}$, so $3 \mid k$ and $9 \mid n$.   □

3   Let $P = A_1 A_2 \cdots A_k$ be a convex polygon in the plane. The vertices $A_1, A_2, \ldots, A_k$ have integral coordinates and lie on a circle. Let S be the area of P. An odd positive integer n is given such that the squares of the side lengths of P are integers divisible by n. Prove that $2S$ is an integer divisible by n. (posed by Russia)

Proof. By Pick's theorem, $2S$ is an integer. We only need to show, for $n = p^t$ a power of an odd prime, $n \mid 2S$.

We prove by induction on k. When $k = 3$, P is a triangle, denote the lengths of its sides by a, b, and c. by the assumption, a^2, b^2, and c^2 are multiples of n. By Heron's formula,

$$16S^2 = 2(a^2b^2 + b^2c^2 + c^2a^2) - (a^4 + b^4 + c^4) \equiv 0 \pmod{n^2}.$$

So $n \mid 2S$.

Now suppose $k \geq 4$, and the proposition holds for all integers less than k. We claim that P has a diagonal whose length squared is divisible by n. If this is true, we can divide P into two polygons P_1 and P_2 by this diagonal and denote their respective areas S_1 and S_2; by inductive hypothesis, both $2S_1$ and $2S_2$ are multiples of n, so is $2S = 2S_1 + 2S_2$.

We prove the claim by contradiction. Assume none of the diagonals of P has length squared divisible by $n = p^t$. Let $\nu_p(N)$ be the p-adic index of N, that is, the highest power of p that divides N. Without loss of generality,

$$\nu_p(A_1 A_m^2) = \alpha = \min_{i,j} \nu_p(A_i A_j^2) < t,$$

where $2 < m < k$. Apply Ptolemy's theorem to the cyclic quadrilateral $A_1 A_{m-1} A_m A_{m+1}$, denote $A_1 A_{m-1} = a$, $A_{m-1} A_m = b$, $A_m A_{m+1} = c$,

$A_{m+1}A_1 = d$, $A_{m-1}A_{m+1} = e$, and $A_1A_m = f$, then $ac + bd = ef$. Square both sides we get

$$a^2c^2 + b^2d^2 + 2abcd = e^2f^2.$$

Since a^2, b^2, c^2, d^2, e^2, and f^2 are all integers, so is $2abcd$. Now we analyze the powers of p on both sides.

$$\nu_p(a^2c^2) = \nu_p(a^2) + \nu_p(c^2) \geq t + \alpha,$$
$$\nu_p(b^2d^2) = \nu_p(b^2) + \nu_p(d^2) \geq t + \alpha,$$
$$\nu_p(2abcd) = \frac{1}{2}(\nu_p(a^2c^2) + \nu_p(b^2d^2)) \geq t + \alpha.$$

So

$$\nu_p(a^2c^2 + b^2d^2 + 2abcd) \geq t + \alpha.$$

On the other hand,

$$\nu_p(e^2f^2) = \nu_p(e^2) + \nu_p(f^2) < t + \alpha,$$

a contradiction. Hence the claim is proved. $\qquad\square$

Second Day

July 12, 2016

9:00–13:30

4   A set of postive integers is called *fragrant* if it contains at least two elements and each of its elements has a prime factor in common with at least one of the other elements. Let $P(n) = n^2 + n + 1$. What is the least possible positive integer value of b such that there exists a non-negative integer a for which the set

$$\{P(a+1), P(a+2), \ldots, P(a+b)\}$$

is fragrant? (posed by Luxembourg)

Solution   The minimum value of b is 6.
    We first prove some facts.

(i) $(P(n), P(n+1)) = 1$.
    This is because

$$(P(n), P(n+1)) = (n^2 + n + 1, (n+1)^2 + (n+1) + 1)$$
$$= (n^2 + n + 1, 2n + 2)$$
$$= (n^2 + n + 1, n + 1) = (1, n + 1) = 1.$$

(ii) $(P(n), P(n+2)) = 7$ when $n \equiv 2 \pmod 7$, otherwise $(P(n), P(n+2)) = 1$.
    Observe that

$$(P(n), P(n+2)) = (n^2 + n + 1, (n+2)^2 + (n+2) + 1)$$
$$= (n^2 + n + 1, 4n + 6) = (n^2 + n + 1, 2n + 3)$$
$$= (4n^2 + 4n + 4, 2n + 3) = (7, 2n + 3).$$

Only when $7 \mid 2n + 3$, i.e., $n \equiv 2 \pmod 7$, $(P(n), P(n+2)) = 7$; otherwise $(P(n), P(n+2)) = 1$.

(iii) $(P(n), P(n+3)) = 3$ when $n \equiv 1 \pmod 3$, otherwise $(P(n), P(n+3)) = 1$.
    By checking modulo 9, when $n \equiv 1 \pmod 3$, $3 \mid n^2 + n + 1$, $9 \nmid n^2 + n + 1$, and $9 \mid 6n + 12$. So,

$$(P(n), P(n+3)) = (n^2 + n + 1, (n+3)^2 + (n+3) + 1)$$
$$= (n^2 + n + 1, 6n + 12) = (n^2 + n + 1, n + 2)$$
$$= (3, n + 2).$$

(iv) $(P(n), P(n+4)) = 19$ when $n \equiv 7 \pmod{19}$, otherwise $(P(n), P(n+4)) = 1$.

$$(P(n), P(n+4)) = (n^2 + n + 1, (n+4)^2 + (n+4) + 1)$$
$$= (n^2 + n + 1, 8n + 16) = (n^2 + n + 1, 2n + 5)$$
$$= (2n^2 + 2n + 4, 2n + 5) = (19, 2n + 5).$$

Only when $19 \mid 2n + 5$, i.e., $n \equiv 7 \pmod{19}$, $(P(n), P(n+4)) = 19$; otherwise $(P(n), P(n+4)) = 1$.

Now we prove that there are no required fragrant sets for $b \leq 5$.

When $b = 2$, $P(a+1)$ and $P(a+2)$ are coprime for any a.

When $b = 3$, $P(a+2)$ is coprime to both $P(a+1)$ and $P(a+3)$.

When $b = 4$, suppose such a fragrant set exists, $P(a+2)$ and $P(a+4)$ are not coprime, so are $P(a+1)$ and $P(a+3)$. This implies $a+1 \equiv a+2 \equiv 2 \pmod{7}$, which is impossible.

When $b = 5$, and suppose such a fragrant set exists, $P(a+3)$ is not coprime to at least one of $P(a+1)$ and $P(a+5)$. It follows that $a+1 \equiv 2 \pmod{7}$ or $a+3 \equiv 2 \pmod{7}$. Now $P(a+2)$ and $P(a+4)$ are coprime, we must have $P(P(a+1), P(a+4)) > 1$ and $P(P(a+2), P(a+5)) > 1$, hence $a+1 \equiv a+2 \equiv 1 \pmod{3}$, which is impossible.

Finally we show that there is a required fragrant set when $b = 6$. By the Chinese remainder theorem, there is an integer a such that

$$a+1 \equiv 1 \pmod{3}, \quad a+2 \equiv 7 \pmod{19}, \quad a+3 \equiv 2 \pmod{7}.$$

Then $P(P(a+1), P(a+4)) = 3$, $P(P(a+2), P(a+6)) = 19$, $P(P(a+3), P(a+5)) = 7$. Therefore $\{P(a+1), P(a+2), \ldots, P(a+6)\}$ is fragrant. $\square$

5 The equation

$$(x-1)(x-2)\cdots(x-2016) = (x-1)(x-2)\cdots(x-2016)$$

is written on the board, with 2016 linear factors on each side. What is the least possible value of k for which it is possible to erase exactly k of these 4032 linear factors so that at least one factor remains on each side and the resulting equation has no real solutions? (posed by Russia)

Solution The answer is 2016.

In order for the equation to have no real roots, the same linear factor can not remain on both sides, so we need to erase at least 2016 of them.

We prove, when we erase all the factors $x - k$, where $k \equiv 2, 3 \pmod 4$, from the left hand side, and erase all the factors $x - m$, where $m \equiv 0, 1 \pmod 4$, from the right hand side, the resulting equation

$$\prod_{j=0}^{503}(x - 4j - 1)(x - 4j - 4) = \prod_{j=0}^{503}(x - 4j - 2)(x - 4j - 3) \qquad \textcircled{1}$$

has no real roots.

We divide cases based on the value of x to prove that the equation $\textcircled{1}$ never holds.

Case 1. $x \in \{1, 2, \ldots, 2016\}$.

In this case, one side of $\textcircled{1}$ is 0 while the other side is not, hence $\textcircled{1}$ does not hold.

Case 2. $x \in (4k + 1, 4k + 2) \cup (4k + 3, 4k + 4)$, where $k \in \{0, 1, \ldots, 503\}$. For any $j \in \{0, 1, \ldots, 503\}$, we have

$$(x - 4j - 1)(x - 4j - 4) > 0,$$
$$(x - 4j - 2)(x - 4j - 3) > 0,$$

when $j \neq k$, and

$$(x - 4j - 1)(x - 4j - 4) < 0,$$
$$(x - 4j - 2)(x - 4j - 3) > 0,$$

when $j = k$. Multiply these together, the left hand side of $\textcircled{1}$ is negative and the right hand side is positive, so $\textcircled{1}$ does not hold.

Case 3. $x < 1$, or $x > 2016$, or $x \in (4k, 4k + 1)$, where $k \in \{1, \ldots, 503\}$. We have

$$0 < (x - 4j - 1)(x - 4j - 4) < (x - 4j - 2)(x - 4j - 3)$$

for any $j \in \{0, 1, \ldots, 503\}$. Multiply together, the left hand side of $\textcircled{1}$ is less than the right hand side, so $\textcircled{1}$ does not hold.

Case 4. $x \in (4k + 2, 4k + 3)$, where $k \in \{0, 1, \ldots, 503\}$. We have

$$0 < (x - 4j + 1)(x - 4j - 2) < (x - 4j)(x - 4j - 1)$$

for all $j \in \{1, 2, \ldots, 503\}$; also $x-1 > x-2 > 0$ and $x-2016 < x-2015 < 0$. Multiply these together we get

$$\prod_{j=0}^{503} (x - 4j - 1)(x - 4j - 4) < \prod_{j=0}^{503} (x - 4j - 2)(x - 4j - 3) < 0.$$

So ① does not hold.

In conclusion, the least number of linear factors we need to erase is 2016. □

6 There are $n \geq 2$ line segments in the plane such that every two segments cross and no three segments meet at a point. Geoff has to choose an endpoint of each segment and place a frog on it facing the other endpoint. Then he will clap his hands $n - 1$ times. Every time he claps, each frog will immediately jump forward to the next intersection point on its segment. Frogs never change the direction of their jumps. Geoff wishes to place the frogs in such a way that no two of them will ever occupy the same intersection point at the same time.

(a) Prove that Geoff can always fulfill his wish if n is odd.

(b) Prove that Geoff can never fulfill his wish if n is even.

(posed by Czech Republic)

Proof. Pick a circle ω that is big enough to contain all the segments in its interior. Extend each segment on both sides until it meet ω at two points. It is clear that we may assume the game is played on these new, extended, segments. So we have n segments as n chords of ω, any two of them intersect inside ω, and no three of them meet at a point. We label the $2n$ endpoints of the n chords as $A_1, A_2, \ldots, A_{2n}$ according to their clockwise order on ω.

(a) Geoff can place the frogs on points $A_1, A_3, \ldots, A_{2n-1}$. First of all, for any chord there are $n - 1$ points on each side of it, so these chords are $A_i A_{i+n}$, $i = 1, 2, \ldots, n$. Thus indeed Geoff placed exactly one frog on each chord. In order to prove that no two frogs will occupy the same intersection, we observe two any frogs, initially placed on A_i and A_{i+2k}, $1 \leq k \leq n/2$, all the indices modulo $2n$. Let P be the intersection of $A_i A_{i+n}$ and $A_{i+2k} A_{i+2k+n}$. It is enough to prove that the segments $A_i P$ and $A_{i+2k} P$ contain different number of intersection points. Each of the chords $A_j A_{j+n}$, where $j = i+1, i+2, \ldots, i+2k-1$ intersects exactly one of $A_i P$ and $A_{i+2k} P$; and every other chord either intersects both A_iP and $A_{i+2k}P$ or none of them. So the total number of intersection points on A_iP and $A_{i+2k}P$ is odd; and we do not have the same number on them.

(b) There must be two adjacent indices A_i and A_{i+1} picked as initial positions by Geoff — otherwise, among the $2n$ positions, Geoff must pick every other position, and since n is even, he must pick some A_i and A_{i+n}, which are on the same chord. Let P be the intersection of A_iA_{i+n} and $A_{i+1}A_{i+1+n}$. Each of the other chords either intersects both A_iP and $A_{i+1}P$, or none of them. So we have the same number of intersection points on A_iP and $A_{i+1}P$; the two frogs initially placed on A_i and A_{i+1} will come to P at the same time. □